Technology in the Policy Process

Controlling Nuclear Power

David Collingridge

 Frances Pinter (Publishers), London

05788/43

First published in Great Britain in 1983 by
Frances Pinter (Publishers) Limited
5 Dryden Street, London WC2E 9NW

British Library Cataloguing in Publication Data

Collingridge,
   Technology in the policy process
   1. Atomic power
   I. Title
   621.48   HD9698
   ISBN 0-86187-319-X

Typeset by Joshua Associates, Oxford
Printed by SRP Ltd., Exeter

D
621·48
COL

To
Ian and Owen

CONTENTS

List of tables   ix
List of figures   xi
List of abbreviations   xiii
Preface   xv

Part I   Nuclear power and theories of policy   1

1   Synoptic rationality and nuclear power   3
2   Incrementalism and nuclear power   23
3   A test for partisan mutual adjustment   41

Part II   The troubled history of nuclear power   53

4   A hypothetical reactor programme   55
5   Nuclear power in Britain   73
6   Nuclear power in the United States   93
7   Nuclear power in France   109
8   Incrementalism and nuclear power again   117

Part III   Tools for control   125

9   Controlling the breeder—flexibility in strategic
       choice   127
10   Special pleading   165
11   Flexibility in tactical choice   191
12   The boundary problem   211

Part IV   Conclusions   227

13   Conclusions   229

Appendix: All you need to know about nuclear power   240

Index   249

LIST OF TABLES

1.1 Factors contributing to LMFBR benefits   7
1.2 Forecasts of electricity demand in the USA   11
1.3 Percentage growth of electricity demand in the USA   11

2.1 Elements of cost for nuclear electricity   31
2.2 Elements of cost for coal generated electricity   32

4.1 Important forecasts in choosing between thermal
    reactor and coal plant programmes   61

7.1 Comparative costs for power plants in France   111

9.1 Elements of cost for breeder electricity   137
9.2 UKAEA breeder reactor scenario   149

10.1 Forecasts for UK primary energy demand 2000   185

11.1 Actual and small-scale LWR programmes   200

LIST OF FIGURES

4.1   Errors in planning the hypothetical reactor
programme are likely   65

4.2   Errors in planning the hypothetical reactor
programme are expensive   67

9.1   Effect of fuel cycle parameters on breeder
doubling time   139

10.1   Comparison of annual world uranium supply and
demand to 2025 low growth projection   167

10.2   Hedging and flexing   173

10.3   Flexining in the Magnox decision   176

10.4   Flexining in the American H-bomb decision   180

10.5   Flexing in breeder design   188

11.1   Reactor experience, actual and $\frac{1}{2}$-size case   201

11.2   Reactor experience, actual and $\frac{1}{3}$-size case   202

A1   Magnox reactor   242

A2   Thermal reactor fuel cycle   244

A3   Breeder reactor   246

LIST OF ABBREVIATIONS

It is unfortunate that no discussion of atomic energy is possible without a peppering of abbreviations. In compensation there are no footnotes in the text. To those who prefer footnotes to abbreviations I apologise.

ABM   Anti ballistic missile.

AGR   Advanced gas reactor.

ʳAPC   Atomic Power Construction—a British nuclear building consortium.

BWR   Boiling water reactor—a variety of light water reactor.

CANDU   Canadian heavy water reactor.

CEA   Commissariat à l'Energie Atomique (French).

CEA   Central Electricity Authority (British), now superseded by the CEGB.

CEGB   Central Electricity Generating Board, generating electricity for England and Wales.

d Old British penny, 240 to the £, about 0.4p.

EDF   Electricité de France, the French electricity utility.

ERDA   Energy Research and Development Administration (USA), now superseded.

GW   Gigawatt, or 10^9 Watts. A generating station with a capacity of 1GW could light up ten million 100-Watt bulbs continuously.

HTGR   High temperature gas reactor.

IAEA   International Atomic Energy Agency.

IDC   Interest during construction.

JCAE   Joint Committee on Atomic Energy (United States Congress).

KW   A kilowatt or 1,000 Watts. The capital costs of power plants are usually expressed in so much per KW of capacity. A plant of 1GW costing £1 billion would therefore have a unit capital cost of £1,000 per KW.

KWhr   A kilowatt hour. The energy provided by one Kilowatt supplied for one hour. Running a one-bar electric fire for one hour provides 1 KWhr of heat.

LMFBR   Liquid metal fast breeder reactor.

LWR   Light water reactor, of which there are two types, the pressurised water reactor and the boiling water reactor.

mbtu   Million British thermal units, a measure of energy.

mill   A thousandth of a US dollar.

MIRV   Multiple independently targeted re-entry vehicle.

mtce   Million tons of coal equivalent.

mtoe   Million tons of oil equivalent.

MW   Megawatt, or 1,000,000 Watts.

NCB   National Coal Board (United Kingdom).

NEC   Net effective cost.

O&M   Operation and maintenance.

ORT   Out of reactor time. The time taken for spent fuel to be returned to the reactor.

PEON   Commission Consultative pour la Production d'Electricité d'Origine Nucléaire.

PWR   Pressurised water reactor, a version of light water reactor.

R & D   Research and development.

SSEB   South of Scotland Electricity Board.

SWU   Separative work unit, a measure of the effort needed to enrich natural uranium.

THORP   Thermal Oxide Reprocessing Plant, at Windscale, England.

TNPG   The Nuclear Power Group, a British consortium for constructing nuclear plants.

TWh   Terawatt hour, equal to one billion kilowatt hours.

UKAEA   United Kingdom Atomic Energy Authority.

USAEC   United States Atomic Energy Commission.

PREFACE

Technologies vary in the ease by which they may be controlled through the normal machinery of politics. Technologies which are difficult to control in this way ought to be avoided. Once embarked upon, they are likely to impose heavy economic and social costs because whatever ill effects they may prove to have, little can be done to alleviate them. This is the theme of this book.

A class of inflexible technologies is identified that are peculiarly difficult to control by the usual processes of policy making. Once such a technology is established, it is impossible to apply any kind of fundamental control for this soon becomes extremely expensive. Whatever ill effects they might have, the only controls which are open are *ad hoc* and superficial. If these controls are ineffective, then those bearing the burden must accommodate themselves to the technology as best they may; there is simply no room for adjusting the technology to their wishes. If political control over technology is to be maintained, if we are going to make proper choices about our technology, if technology is to serve human interests, and if human wishes are not to be curbed by technology, then inflexible technology must be avoided, for it forces us to fit in with its own imperatives and cannot be bent to suit our own, merely human, purposes. If in places I have been unable to make a point without arid technicalities, may I be forgiven, for my theme is deeply serious and pressing—it is no less than human freedom.

The discussion is built around nuclear power, although I hope that the argument can be followed by those with only a minimal acquaintance with the technology itself, whose understanding may be aided by the appendix. The technology of nuclear power is shown to have features which make it very inflexible in the sense that, once built, it is very difficult and expensive to control. These features are its high capital cost and capital intensity, long lead time, large unit size and its dependence on specialised infrastructure for its working. These combine to ensure that whatever problems the technology produces, be they economic, social or political, there is little that can be done to ease them. Anything more than *ad hoc* controls are bound to be very expensive and slow to work. If political control over technology is to be kept, then nuclear power must be avoided. This is no real loss, however, because the same inflexibility which shields nuclear power from political control also makes it a very risky and troublesome investment, through making learning about the technology both slow and expensive. To put it another way, the inflexibility of nuclear technology means that errors are likely to be made in its planning, and that whatever mistakes are made are likely to prove very costly. There are, therefore, two ways of expressing the case against nuclear power: the technology cannot be subject to political control; and its development involves far greater risks than ordinary technological innovations. This theme is developed through a historical discussion of nuclear power in Britain, the United States and France.

If inflexible technology is to be avoided, then it is clearly crucial to be able to identify a technology that is inflexible at an early stage in its development, before it has acquired an immunity to political control. Again, this problem is approached through the example of nuclear power, in particular the breeder reactor. It is shown that a technology may be identified as inflexible on very little information, on far less than is needed for a calculation of its likely costs and benefits, so that considerations of flexibility may enter the policy making process very early, long before the technology has developed a resistance to control.

The breeder is shown to be even less flexible than today's nuclear technology because it will have higher capital costs, be of greater capital intensity, need a longer lead time and a larger unit size and will require more infrastructure for its operation. If it is developed, the breeder will therefore be even less open to political control than the nuclear plant of the present; its planning will be more prone to errors, and whatever errors are made will be more costly than for existing nuclear technology. It is therefore even less socially and economically acceptable than today's nuclear power, despite the central place which the breeder continues to have in so much energy planning.

It gives me great pleasure to thank all those friends, colleagues and students who have contributed so much to my thinking, and to the Technology Policy Unit as a whole for providing such a congenial atmosphere for this kind of work, and to Gill Crawford for a first class piece of typing. The last word of this book was written on the day which also saw the departure of the Unit's first Director, Professor Ernest Braun. May whatever is of value in here be my farewell to him. Finally, I wish to express my gratitude to my wife Jenny for her support in so many ways in the writing of this work, and so much more.

David Collingridge

PART I

NUCLEAR POWER AND THEORIES OF POLICY

1 SYNOPTIC RATIONALITY AND NUCLEAR POWER

The concern of this book is theories of policy making and their ability to accommodate decisions about large technological projects, as an example of which nuclear power is taken for purposes of illustration. The natural starting place for any discussion of theories of policy making is the view of *synoptic rationality*, if only because all other views have been developed as a reaction against this one. Synoptic rationality demands that a decision maker *justify* whatever choices he may make, a task which involves no mean effort. A policy maker is supposed to identify which of his values is relevant to the problem in hand, and to undertake a comprehensive survey of all possible options which might further these values. All the consequences, or likely consequences, of adopting each of these options is to be exhaustively listed, so that a decision can finally be made which maximises the attainment of the policy maker's values. There are variations on this theme: some insisting that cost–benefit calculations be done, some that cardinal utility be explored; some requiring maximisation of benefit, others working with the weaker idea of satisficing. But there is more uniting these variations than separating them, in particular the demand that choice be justified through a comprehensive review of relevant values and of options and their likely consequences (Carley (1980), pp. 11 ff. and Friedmann and Hudson (1974)). In practice, such a synoptic view can be achieved only for humble decisions, such as those involved in playing simple games. Collingridge (1982) contains an extended discussion of the scope of any kind of decision theory which insists that choice is justified. Nevertheless, synoptic rationality is appealed to as an ideal to which policy makers should aspire, even if it is impossible to reach in the cruel world they inhabit (Banfield (1959), Braybrooke and Lindblom (1963), Meyerson and Banfield (1955) and Meyerson (1956)).

This chapter will consider whether decisions about large-scale technology can be made according to the synoptic ideal. A discussion of this question can usefully be based around the cost–benefit analysis of the liquid metal fast breeder reactor (LMFBR) undertaken by the United States Atomic Energy Commission (USAEC). The version to be considered here was published in 1974 (USAEC (1974)), and in slightly revised form (ERDA (1975)), it was submitted to the Environmental Protection Agency as part of the Environmental Impact Statement on the breeder reactor. It was preceded by earlier versions in 1968 and 1970 (USAEC (1968) and (1972)), these being amended in the light of comments from other government agencies and the general public. Throughout this period USAEC saw the breeder as the logical next step in reactor technology, and was very anxious to aid its development by an intensive R & D programme including the building of a large-scale prototype, after which it was hoped the LMFBR would become fully commercial. The discussion of the cost–benefit analysis which follows is understandable without a detailed knowledge of the 'nuts and bolts' of the technology, but some readers may wish first to look at the appendix, where a brief description of the technology is presented.

The cost of the LMFBR R & D programme was estimated to be around $5 billion (like all figures from the cost–benefit analysis, this is in 1974 US dollars; where other years are used, this will be noted). The problem for USAEC, therefore, was to show that the benefits from the LMFBR would be very likely to exceed these development costs, showing that their favoured reactor was a good investment. This is the aim of their cost–benefit analysis. It is important to remember that all the development costs would have to be borne before the reactor will begin to show any benefits; if, for whatever reason, the reactor should fail to deliver the benefits which are expected of it, there is no way of saving its R & D costs because these have already been sunk. The commitment of such large R & D resources therefore calls for a high degree of certainty that the reactor will deliver adequate benefits. This is why the synoptic view of the cost–benefit analysis was felt to have been needed. That the cost–benefit analysis really does require a synoptic view of the breeder reactor will soon become apparent from a brief description of the study.

The benefits from the breeder reactor were taken as the cost of supplying electricity demand from 1974 to 2020 (the study's time horizon) *without* the breeder, minus the cost of supplying electricity demand from 1974 to 2020 *with* the breeder. This is logically impeccable, but the simple subtraction disguises the true complexity of the calculation. First of all, of course, electricity demand in the United States from 1974–2020 needs to be known. This straightaway highlights the first central difficulty of the analysis; benefits are acquired from the breeder so late that the analysis needs to take a very long look into the future, some forty or fifty years, with the attendant hazards of forecasting over this span of time. Just how great these problems are will be judged later. The cost of generating electricity from breeder reactors also needs to be known over the same period. Here we come across the second principal difficulty faced by those undertaking the analysis; the breeder reactor is so very different from rival reactors and from fossil plants, that comparing the costs of their electricity requires knowledge of a whole range of factors. The cost of electricity from the breeder depends upon its capital cost and its fuel costs (ignoring the much smaller O & M costs). Capital costs will, of course, be sunk so that their contribution to the cost of a unit of electricity will depend upon the reactor's load factor, or intensity of use. So capital costs and load factor for the new reactor need to be known, but these may vary over time, and they need to be known from 1974 to 2020. The breeder's fuel costs depend upon the cost of reprocessing, transporting, storing and fabricating its fuel and of managing the wastes derived from the fuel. These costs therefore need to be known and, as before, each will have a fixed and variable component, since special reprocessing, storing, fabrication and waste treatment plants will have to be constructed. As before, the contribution of the fixed costs to the cost of one unit of electricity will depend upon the total quantity of electricity generated by the breeder.

To find the cost of meeting electricity demand without the breeder, the make-up of the cheapest generating system that has no breeder needs to be known. This is dominated by thermal reactors according to the forecasts of the analysis, but there are two very different varieties to choose between, light water (LWR) and high temperature gas (HTGR) reactors. This choice introduces a considerable complication: not only will the two reactors have quite different capital costs, which can be expected to vary differently over time, but they have quite separate fuel cycles. Comparison of their fuel costs

therefore involves detailed knowledge of the costs of the components of each fuel cycle, the fabrication, reprocessing, storage,
waste plants, etc. Again, these may be expected to vary differently over time. Nevertheless, the analysis offers forecasts
for these costs, and this enables the cheapest way of meeting
electricity demand over the study period, where the nuclear
component of the generating system is split between LWRs
and HTGRs, to be determined. This cost may then be compared with that of providing the same electricity where breeder
reactors are used as well.

Using the forecasts thought by the analysts to constitute
the most likely future, their 'base case', and discounting future
costs and benefits to obtain a net present value, the report
claims that at a discount rate of 7 per cent development costs
of the LMFBR are $5 billion, and the net present value of
the proposed reactor is reckoned at $54 billion. At a 10 per
cent discount rate these become $5 and $19 billion respectively, development costs not being sensitive to discount rate
as they arise over a relatively short period. Table 1.1 gives
a summary of the major benefits to be derived from the
breeder with a 7 per cent discount rate. The chief benefits
are a reduction in the demand for uranium, from a cumulative
6.3m tonnes by 2020 to 2.6m tonnes, and the work necessary
for its enrichment. By 2020, 450 million separative work units
(SWU) per year would be needed without the breeder, as
against a peak of 95 million SWU per year in 2015 with the
breeder.

There are, of course, considerable uncertainties in any
analysis which depends on such a variety of long-term forecasts, therefore sensitivity tests were performed. Values of
the parameters whose future values were uncertain were adjusted one way or the other from their value in the base case
to see what effect this would have on the net benefits of the
breeder reactor. Thus, for example, three relationships between
the cumulative supply of uranium and uranium mining costs
were used, and four forecasts for electricity demand. In all,
seventy-two cases were considered. It was claimed that in
nearly all of these cases, the benefits from the breeder would
exceed the cost of its development. Costs are greater than
benefits 'only when two or more large adverse circumstances—
each considered unlikely—are assumed to occur in concert'

Table 1.1. Factors Contributing to LMRBR Benefits*

	Net benefit ($ 10^9)
Plant capital	−2.8
Plant operation	−3.7
Purchase of uranium	+43.9
Separative work	+16.3
Others	+0.2
Total benefit	+53.9

*Costs are negative, benefits positive.
Source: USAEC (1974), Table 11.2–19.

(USAEC (1974), 11.3). The report concludes that breeder reactors offer the promise of a nuclear industry independent of uranium supplies, and a generating system whose costs decline over time. 'The fast breeder may well be the doorway to unlimited low-cost power which can enhance the Nation's prosperity' (USAEC (1974), 11.3), and it should therefore be introduced as quickly as possible.

The analysis also gives a great deal of attention to the effects the breeder would have on man and on the environment. The conclusion of these studies is, however, that the effects of the breeder would be substantially the same as the effects of the next best electricity system dominated by LWRs and HTGRs. This is a happy accident, there is nothing about the reactors and their fuel cycles which guarantees such an outcome. There was therefore no short cut to the conclusion that environmental effects are roughly equivalent; this was determined only after what attempted to be a synoptic view of the environmental impact of each reactor.

So far, enough has been said about the USAEC's cost–benefit analysis to show that it attempts to achieve a synoptic view which captures all the costs and benefits of the breeder reactor and its alternatives. If this is accepted, we may pass on to consider just how realistic this aim of a synoptic view is. It is now time to raise the whole question of the usefulness of striving for a synoptic view in the case of a technology like the breeder reactor.

The central obstacle to achieving a synoptic view of the breeder reactor is that its introduction would be a *non-incremental* technical change. What this means may best be grasped by considering the other end of the spectrum, *incremental* technical changes. When there is only a minor difference between the existing state of affairs with which we are familiar and a proposed change, or between two proposed changes, then comparison is easy because attention may be focused on the difference between the proposals or between the status quo and the proposed change. Consider, for example, the problem of tackling lead pollution from motor vehicles. If this is to be done by reducing the legal limit of lead in petrol, then moving from a limit of, say, 0.84 g/l to one of, say, 0.64 g/l buys a calculable reduction in air lead concentrations at a cost to the refiner which can also be determined. These benefits and costs may then be compared with those for other limits. A reduction of the limit of lead in petrol is therefore an incremental change. There is no need to have detailed knowledge of the fate of lead from vehicle exhausts, or of the workings of the petrol engine, nor to forecast future mileage, or air lead concentrations. The policy maker may be presented with a number of options giving benefits in terms of lower concentrations of lead in the air and costs in money from which a choice may be made. But now suppose it is proposed to tackle the problem not by a reduction in lead levels in petrol, but by a major shift from petrol-driven motor cars to (lead free) diesel-fuelled buses and trains. Trying to assess such a change opens up an enormous number of questions. Will people resist such an attempt? Should people who have invested in motor cars be compensated for restrictions on their use, and if so, how? Will enough diesel oil be available from the refineries in the time demanded for the shift to public transport? What can be done with the surplus petrol provided by the refineries? What will the rapid expansion of rail and bus services cost? How quickly can manpower for the new public services be recruited and trained? And so on and so on. In short, the proposed shift to public transport is a non-incremental one: estimating its costs and benefits involves giving attention to a whole set of consequences.

To give a second example nearer home, comparing a 500-MW coal-powered generating plant with two 250-MW coal-powered plants of similar design is relatively straightforward. The larger plant can be expected to be more efficient, so that there will be a saving of coal; the capital cost will probably be less than that for the two smaller plants together, but the larger plant is likely to take longer to build. In comparing the two, analytical attention can be focused on these well defined differences. There is therefore an incremental difference between them. But it is quite otherwise for a coal station and a nuclear station of the same output. Calculation of the difference in cost of producing electricity from coal and from uranium in, say, an LWR requires detailed knowledge of the entire costs of each method of generation. Capital costs and load factors for the coal plant need to be estimated, and coal costs need to be forecast for the period of its operating life. To these direct costs need to be added the indirect costs of the environmental damage of winning the necessary coal, and the impairment to health from coal mining and from breathing air polluted by sulphur dioxide and soot. On the nuclear side, the capital costs and load factor of the LWR need to be forecast, as does the price of uranium and the cost of the metal's enrichment. Allowance must be made for reprocessing benefits (if any), the sale of recovered plutonium (if any), the management of the wastes from the reactor and its final decommissioning. To these direct costs must be added the indirect costs arising from the effects of the normal operation of the plant on the environment and the plant's operators, and the effects of any accidents which may happen. Indirect costs from waste disposal and uranium mining must also be included in the calculation. Comparing chalk and cheese seems pretty straightforward in the face of these problems. Considering a coal-fired plant against an LWR, or any sort of nuclear reactor, is clearly a non-incremental comparison demanding very large quantities of information about a whole range of items and many long-term forecasts.

The USAEC's cost–benefit analysis tries not only to compare the costs of generating electricity by breeder reactors and by coal-fired plant, but also tries to estimate the cost of generation by two quite different non-breeding reactors, LWRs and HTGRs. Comparing these very different systems requires an enormous amount of information, leading directly to two problems which beset any attempt to gain a synoptic view of a non-incremental change: the problem of *forecasting* and the problem of *entrenchment*. Much of the information which is required concerns the future and forecasts of very many factors must be made, forecasts which are inherently uncertain. This is the first problem, and it is worsened by the long lead time of the breeder reactor and its associated plants, forecasts over fifty years often being required. The second problem arises because the synoptic view is only an ideal; in practice it can be aimed for, but there will always be uncertainty about some of the information required by a synoptic view. In the present case, such a quantity of information is required that many items are, at the time of the study, highly uncertain. It may therefore happen that by the time future research has eliminated the uncertainty surrounding a particular item of information, in such a way that the information counts against the breeder reactor, the technology will have become so embedded in the whole economic fabric that changing it is hugely expensive and slow, even to the point of being impossible. This is the problem of entrenchment. In other words, by the time information arrives, it may be impossible to use it to control the technology. The two problems will be discussed separately.

1. The forecasting problem

Let us consider the forecasting problem by way of some examples from the USAEC's cost–benefit analysis.

(a) Electricity demand and nuclear component

Electricity demand is forecast exogenously. In 1974 US electricity demand was 1.7×10^{12} KWhrs and was increasing at 6.8 per cent per annum. Table 1.2 shows the four forecasts used in the cost–benefit analysis.

The accuracy of some of these forecasts can be assessed with the data now available (November 1982). Table 1.3 shows the real growth of electricity production from 1974 to 1980.

History speaks loudly against all of the four electricity demand forecasts. All four differ only slightly in their forecasts up to 1980, as they all start at 7.8 per cent for 1974, declining at different rates to 2020. The base case percentage increase in 1980 is 7.32 per cent, the large case giving 7.41 per

Table 1.2. Forecasts of Electricity Demand in the USA

	Growth rate of electricity demand		Demand in 2020 (10^{12}kWhr)
	1974 (%)	2020 (%)	
Base	7.8 steadily declining to	3.7	28
Large	7.8	4.2	34
Small	7.8	3.0	22
Very small	7.8	2.8	14

Source: USAEC (1974), 11.2.1.2.

Table 1.3. Percentage Growth Rate of Electricity Demand in the USA

1974	0.1
1975	1.8
1976	6.1
1977	4.1
1978	3.3
1979	1.4
1980	1.6

Sources: UN Yearbook of World Energy Statistics (1980); *Statistical Abstracts of the United States* (1979).

cent, the small 7.20 per cent and the very small 7.14 per cent. The spread is only 0.3 per cent. But the real growth rate in all years up to and including 1980 was far less than even the very low forecast. Even in its first year, 1974, the forecast is very poor, 7.8 per cent being calculated, which in reality came to 0.1 per cent. The seven years' experience clearly show that the forecasts of demand in the USAEC study are far too high.

(b) Capital costs

Capital costs for LWRs are taken as $420/KW from 1970 to 1990, after which an increase in the size of unit from 1.3 to 2.0 GW brings about a reduction to $368/KW. In the absence of operating experience of commercial HTGRs, the capital costs of this reactor are simply assumed to be the same as those for the LWR, and to show the same reduction after 1990. The initial capital costs of the LMFBR are reckoned to be $520/KW, $100/KW more than competing reactors. Learning and serial production are then supposed to reduce this differential linearly to $70/KW by 1990. After this, the unit size of breeder reactors increases from 1.3–2.0 GW and the differential in capital costs moves linearly to zero by 2000. To cope with the uncertainties in forecasting capital costs for three reactor types so far into the future, two variations on this most probable, or base case, are considered. In the first, the capital costs of breeder reactors are still $50/KW above their rivals' costs by 2000, and remain so until the end of the study period in 2020. The second variation makes the cost differential $100/KW by 2000 and beyond.

The whole tangled history of nuclear reactor capital costs will be discussed in detail in later chapters. For the present it may be enough to show that assumptions embodied in this part of the USAEC's study run counter to nearly twenty years' experience in constructing nuclear plant. The study, as we have seen, assumes that LMFBR, LWR and HTGR capital costs will remain constant. History shows that capital costs have consistently risen since the earliest days of atomic power, and are at present doing so at something like the rate of $ (1978) 140/KW/yr. The study assumes that all three nuclear plants will exhibit scale economies, enabling considerable reductions in their capital costs to be achieved. To the present, however, it is doubtful whether there are any scale economies, and if there should be they are very much less than those assumed here. The LMFBR is supposed to follow a learning curve, with decreasing capital costs, and yet there are good reasons to expect that any new type of reactor will show just the opposite. As operating experience reveals more and more ways in which the reactor might fail, there is a relentless increase in the number of failure modes which must be safeguarded, and so capital costs rise instead of fall. This appears to be true of LWR designs, of which there is the greatest experience, and it can be expected to be true of any future design of reactor. When the study's forecasts are compared with history in this way it is quite apparent that its basic assumptions about capital costs are in error, and that the attempt to cope with uncertainty by considering the two very modest variations on the base case is trifling. The least which could hope to be called reasonable would be variations with increasing capital costs over time and no economies of scale. These would, of course, be very unfavourable to the breeder reactor.

(c) *Uranium prices*

The report's treatment of uranium prices casts an interesting light on the problem of trying to draw practical bounds in an attempt to gain a synoptic view of a problem. The study makes a set of assumptions which together ensure that only uranium mined in the United States needs to be considered in the cost–benefit analysis. It is supposed that there is a real world-wide market for uranium, that there is no price fixing by cartels and that the United States would be self-sufficient in uranium. Under these circumstances, the problem reduces from the world-wide supply of metal to supply from sources within the United States, so that it becomes much more manageable. Three estimates are made of the quantity of American uranium which could be extracted before the more expensive shale deposits need to be exploited. The base case is 4 million tons, with a high case of 6 and a low of $2\frac{1}{2}$ million tons. With other assumptions about marginal resources and the effort put into exploration for new reserves, a relationship can be derived between cumulative demand for uranium and the metal's marginal price for the three cases. The three cases are needed, as before, to try to cover the great uncertainty which surrounds knowledge of uranium resources, and they are used in sensitivity testing in the cost–benefit analysis.

In the report it is stated: 'the USAEC fully appreciates that the pivotal question *vis-à-vis* the rapid development of the breeders is the prospect for finding additional uranium resources with grades of ore that are higher than that in the shales' (USAEC (1974), 11.2.3.7). In short, if there is plenty of uranium in the ground which can be easily mined, then existing reactors will produce cheaper electricity than the breeder. If this easily available uranium is scarce, then conversely, the breeder will be the favoured source of electricity. But the above remark and the USAEC's cost–benefit analysis, consider only resources within the United States. The report's authors, recognising the key nature of this question for the breeder, point to surveys under way in the United States which will eventually produce a national appraisal of uranium resources.

But is this restriction to America a realistic one? Consider a case where the national appraisal is pessimistic and discovers few resources better than the shales, so that investment is made in breeder reactors, but uranium reserves outside the United States are found to be very cheap to exploit. The United States would have an electricity system independent of imported uranium, its few thermal reactors running on its little indigenous uranium, and its many breeders not requiring uranium. But the cost of independence could be very high indeed. Electricity might be produced much more cheaply using imported uranium and mostly thermal reactors. Whether the savings would justify the loss of control over fuel is a question which has to be left to the politics of the future. The attempt to confine the question of uranium prices to United States' uranium is therefore a dangerous oversimplification, which could prove to be very expensive. Attention must be paid to the world-wide availability of the metal and its world price. This makes a great deal of work for the analysts because not only must an estimate be made of the world-wide distribution of winnable uranium, but its price must be calculated. This depends, of course, on demand for the metal, which in turn depends upon the rate at which uranium-burning reactors are built across the world, which in turn depends upon energy demand across the world. The restriction of the uranium question to the United States means that the cost–benefit analysis can confine itself to United States' energy demand and consider only the penetration of thermal reactors in that country. Once it is realised that world-wide uranium resources need to be considered, then so too must world-wide energy demand, electricity demand and demand for uranium. This is a vastly expanded analytical problem. It has been tackled by braver analysts than those used by the USAEC (World Energy Conference (1978)), but the kind of assumptions which have had to be made are really quite untenable. Thus the attempt to confine the study to the United States fails, the world should be the topic of study. This often happens when a synoptic view is attempted; it proves impossible to bound the analysis, and its scope inevitably expands outwards until everything is encompassed. This *boundary problem* is a matter for discussion in chapter 12.

   Later chapters will look at the availability of uranium, but for now all we need to note is that estimates of world-wide uranium resources have consistently increased over recent years, and the fears of uranium shortages that were once so widespread have gradually subsided. In this light, the USAEC's forecasts of uranium reserves seem much too pessimistic.

The first central problem of any attempt to obtain a synoptic view of a proposed technical change is the demand which is placed on forecasts. Forecasts of a whole number of things are needed if a synoptic view is to be achieved, and in the present case the problems of prediction are compounded by the very long time it takes the breeder reactor to gather benefits which might outweigh its development costs. We have seen that three key forecasts—of electricity demand, capital costs and uranium prices— are all in error to a significant degree; and this by 1982. What the difference between reality and forecasts might amount to by 2020 will one day be an interesting exercise. Other forecasts made in the study could equally well have served our purpose. Indeed even the short-term forecast of the breeder's development costs has been severely criticised as perhaps three times too low (Alexander and Rice (1975)).

This is the problem, but in which direction lies the solution? One suggestion is that the art of forecasting be improved until confidence can be felt even in the multiple long-term forecasts required to achieve a synoptic view of the breeder reactor. But even the greatest of optimists must admit that such learning will be acquired only very slowly. An essential in learning how to forecast is the comparison of the forecast with reality when that time in the future is eventually reached; but what if the forecast is trying to probe thirty or forty years ahead or more? Learning must wait many years for such forecasts. Consider for example, the forecast of electricity demand from 1974 to 2020 discussed above. The forecast is wrong in its early years, as we have seen, but is this an error which vitiates the whole forecast, or one which will appear as a mere short-term deviation when 2020 finally arrives? There is no way of telling. Learning that the forecast is substantially right or wrong will take many more years. Thus learning about the success or failure of forecasts and methods for their making is a very long-term enterprise. Before we have the blessing of such knowledge, we need to find ways of coping with our inability to forecast with any degree of certainty. Those of a pessimistic cast of mind, like myself, would see this inability as a permanent feature of man's lot. Either way, methods of coping must be found, either temporarily or permanently (Collingridge (1980)). This is really the purpose of the present work: how can we adjust our choice of technology to accommodate the very limited powers of prediction with which we are blessed? It may be that learning to live with our ignorance about the future entails learning to live without the breeder reactor.

2. The entrenchment problem

The cost–benefit analysis presented by the USAEC attempts to take a synoptic view of the introduction of the breeder reactor. It tries to identify all the differences which the new technology would make to the world, not just the narrow economic losses and gains, but also the environmental effects of the breeder and of its rival sources of power. This, as we have seen, involves a very wide-ranging study involving many long-term forecasts of doubtful credibility. The second problem with this approach to making decisions about the breeder stems from the uncertainties which inevitably surround many of the issues relevant to the choice between different sources of electric power. Uncertainties are bound to exist in any study approaching the magnitude of the present one, and this obscures the synoptic view. Nevertheless, analysts must do their best, and taking the synoptic view as an ideal to be aimed at they naturally try to reduce the uncertainties as much as possible, where the data required are available, and to use sensitivity tests where this cannot be done. The analyst must do his best, must make whatever judgement of the facts seems at the moment to be most reasonable and defensible. If the synoptic ideal is unobtainable, it can nevertheless be approached in this way. If certainty about all relevant information is impossible, we must adjust to this by working on those hypotheses which seem most reasonable at the moment. This simple rule is, however, a trap. This is nicely illustrated by the study's treatment of the toxicity of plutonium.

The study recognises the need for concern about the health hazards posed by the breeder's production of transuranic elements, and particularly plutonium. Section 1.4.3 tells us that

(1) large quantities will be produced;

(2) the elements are very long lived;

(3) there is no direct knowledge of their effects on man;

(4) such effects as might conceivably occur might be indistinguishable from the normal ills of mankind.

The study estimates the total emissions of plutonium and other transuranics from the normal operation of the breeder's fuel cycle, and from accidents at various places in the cycle. The fraction of these which finds its way to man is then estimated, and the subsequent distribution of this to various organs is calculated. Of these results, the study states that

Although a substantial data base exists, some of the information required to make precise predictions is not available. In the absence of complete data, conservative estimates have been made using the best available information. Therefore, there is a high probability that estimated potential health consequences are overestimated. [USAEC (1974), 4.7 appendix II.G]

Even with this cautious approach, the study concludes that normal releases from the breeder's fuel cycle by 2020 will be less by a factor of 20,000 than the plutonium deposited from American bomb tests; that the inhalation of transuranic elements from the fuel cycle will deliver a dose of α particles only one-ten thousandth of that from natural sources; and that doses to lung, bone and gonads will be lower than naturally occurring doses by a factor of between 10^4 and 10^6. With such results, plutonium toxicity does not seem to present a serious impediment to the breeder's development, as indeed the study argues.

In doing these sums, however, many simplifying assumptions have had to be made. For our present purposes it will suffice to pick out one of these: that the dose to the lung is evenly distributed throughout the organ. This simplifies calculations because there is no way of predicting the size distribution of the radioactive particles reaching the lung. There have been suggestions that this simplification seriously underestimates the damage done to the lung because particles of a particular size ('hot particles') might damage the lung very much more seriously than the same dose of radiation spread evenly throughout the volume of the lung. The hot particle hypothesis is discussed in the study. Theory is recognised as being quite inadequate for judging the hypothesis, as are the small number of animal experiments which have been done to date. There are, however, many people who have been exposed to airborne plutonium and other transuranic elements in the course of their work, although in only one case, the fire at Rocky Flats, was any attempt made to measure the particle size distribution. If the hot particle hypothesis were correct, then excess deaths should have resulted in workers exposed by the Rocky Flats fire, and yet none were found. If a similar particle size distribution is assumed for other cases, then there should have been 5,000 deaths among workers exposed to airborne plutonium. This has not occurred, and so the hot particle hypothesis is rejected.

The case made out against the hot particle hypothesis is far from complete. The uncertainties surrounding the hypothesis are clearly appreciable, and some version might be found to be true by future researchers. If today's best judgements on such technical matters are to be used in the decision whether to develop the breeder reactor, then the hot particle hypothesis should be put aside. The best judgement of the experts of 1974 was that plutonium toxicity could be ignored in the balancing of the breeder's costs and benefits, it is just not an important factor. But is this a reasonable way to make decisions about technology? Is it reasonable, that is, to decide what technology to adopt by referring simply to the best judgements which can be made at the time, even though these fall short of certainty and may be radically revised in the light of research work yet to be done? Today's best judgements need to be supplemented by some assessment of the costs which would be incurred if these judgements were later revealed to be in error. Only then can a rational assessment of the technology be made.

Suppose, then, the LMFBR is developed and in 2000 it is discovered that plutonium is far more toxic than the best estimates on the data available in 1974 would have. If the forecasts of the USAEC study are to be believed, then by this time a considerable fraction of the United States' electricity would be provided by breeder reactors, say a half. Electricity cannot be substituted for in the short term because this requires the scrapping of electric plant and the buying of plant running on some other source of energy. No matter how severe the dangers from plutonium might be found to be, an immediate halt to generating electricity from the breeder reactors would be hugely damaging to the economy, especially since electricity is an expensive fuel and is used for premium purposes. Supply could not be shifted to other generating plants because there are not enough of them, nor would there be enough coal and uranium being mined to fire them even if there were. Construction of alternative stations would take many years to complete because it takes seven to twelve years to build a single large plant, and a crash programme would soon run into the bottleneck of limited construction capacity, which would itself take many years to resolve. By the time it has been discovered that plutonium presents a much worse hazard than anyone thought in 1974, the breeder reactor has become essential to the operation of the whole economy, and can only be abandoned in the short term at huge cost. Many steps might be taken to alleviate the problem: load-following coal stations might be used for base load plant, coal being won from easily started strip mines and shallow mines and so on, but these could not hope to make much impression. It might be, of course, that the newly discovered toxicity of plutonium would not merit this drastic abandonment of breeder reactors, but could be handled by marginal improvements in controlling emissions from plants. But the reverse could also be true; the problem could be found to be so severe that more than marginal changes are required for its control, and these will then be very expensive. We may say that the breeder will have become *entrenched* (Collingridge (1979), (1980), chapter 3). The whole economy will have become adjusted to the provision of cheap and plentiful electricity from breeder reactors, so that any attempt to impose radical and quickly operating controls on the technology will have become hugely expensive. Entrenchment means that the cost of today's best guesses being wrong can be enormous. In deciding therefore what technology to adopt not only must the best guesses about technical issues be made, the cost of later finding them to be wrong must also be brought into the account. In the case of the breeder and plutonium toxicity, this counts heavily against the breeder; any mistakes could be extremely expensive if they are discovered after a large part of the economy has come to depend on the breeder.

Could these difficulties be met by not going ahead with the breeder until we are certain of the technical issues? Should the breeder's development be postponed until, for example, there is certainty that plutonium is not much more toxic than we think at present? This would avoid the problems caused by the technology's entrenchment, but at a massive cost. Technical issues like that of the toxicity of plutonium often take many years to be resolved and so postponement of the breeder project until all these issues have been settled would mean postponing the technology's benefits for thirty to forty years or more. The very novelty of the technology means that there are a host of technical issues surrounded by uncertainty, and presumably this suggestion would mean delaying until all of those are answered—a daunting prospect.

Thus it can be seen that the time to be required to eliminate the uncertainty surrounding plutonium toxicity places the breeder's proponents in a dilemma. The reactor can be developed before there is a clear picture of plutonium's effects on the health of the population, in which case there is the risk of discovering that plutonium is more toxic than presently thought at a time so late in the breeder's development that little room for control exists. On the other hand, the breeder might be delayed until the issue of plutonium toxicity is settled, but this is sure to postpone whatever benefits the breeder might bring for several decades. Animal experiments are of very little relevance and it is ethically impossible to experiment on humans. Understanding has therefore to wait upon accidents like the Rocky Flats fire. Even then, it may take many years for health effects to be revealed, and the picture will be muddied if these effects resemble those of common diseases. The problem caused by plutonium toxicity may, of course, be generalised. The very novelty of the breeder ensures that there are many technical issues of similar uncertainty concerning for example reactor safety, the safety of reprocessing plants, the infringement of civil liberties necessary to safeguard the technology against terrorist attack and the encouragement the breeder will give to the proliferation of nuclear weapons. These will be discussed more fully later, but the point to be made here is that they pose the same dilemma as that surrounding the toxicity of plutonium.

In conclusion, it may be said that USAEC's brave attempt to follow the synoptic ideal of decision making in its cost–benefit analysis of the breeder is a failure. The ideal of synoptic rationality is wholly inappropriate for making decisions about technologies like the breeder reactor. Many long-term forecasts need to be made, but they cannot be made with the required degree of accuracy. This is not a reflection on the forecasters themselves because learning about long-range forecasting is essentially a very slow business. Secondly, many technical uncertainties remain. If the breeder's development is delayed until research has settled all of these issues, then the technology's benefits will be greatly delayed. If, on the other hand, a decision about the breeder is made on the best judgement which can be made today, then some of these opinions are likely to be discovered to be in error, but at a stage in the breeder's development when there is little room to respond to this revelation. We must, it seems, look for some other way of making decisions about technology than synoptic rationality.

References

Alexander, A. and D. Rice (1975), *Comments on the LMFBR Cost Benefit Analysis*, RAND Corporation, AD–AO22–296, Santa Monica.

Banfield, E. (1959), 'Ends and Means in Planning', *International Social Sciences Journal,* **9**, reprinted A. Faludi (ed.) (1973), *A Reader in Planning Theory*, Free Press, Glencoe, pp. 139–49.

Braybrooke, D. and C. Lindblom (1963), *A Strategy of Decision*, Collier-Macmillan, London.

Carley, M. (1980). *Rational Techniques in Policy Making*, Heinemann, London.

Collingridge, D. (1979), 'The Entrenchment of Technology', *Science and Public Policy*, **6**, 332–8.

Collingridge, D. (1980), *The Social Control of Technology*, Frances Pinter, London.

Collingridge, D. (1982), *Critical Decision Making*, Frances Pinter, London.

Friedmann, J. and B. Hudson (1974), 'Knowledge and Action', *Journal of the American Institute of Planners,* **40**, 2–16.

Meyerson, M. (1956), 'Building the Middle-Range Bridge for Comprehensive Planning', *Journal of the American Institute of Planners,* **22**, reprinted A. Faludi (ed.) (1973), *A Reader in Planning Theory*, Free Press, Glencoe, pp. 127–38.

Meyerson, M. and E. Banfield (1955), *Politics, Planning and the Public Interest*, Free Press, Glencoe.

United States Atomic Energy Commission (USAEC) (1968), *Cost Benefit Analysis of US Breeder Reactor Program*, WASH–1011, Washington D.C.

United States Atomic Energy Commission (USAEC) (1972), *Cost Benefit Analysis of US Breeder Reactor Program, updated 1970*, WASH—1184, Washington D.C.

United States Atomic Energy Commission (USAEC) (1974), *Liquid Metal Fast Breeder Reactor Program, Environmental Statement*, WASH-1535, Washington D.C.

United States Energy Research and Development Administration (ERDA) (1975), *Liquid Metal Fast Breeder Reactor Program, Final Environmental Statement*, ERDA-1535, Washington D.C.

World Energy Conference (1978), *World Energy Resources 1985-2020: Nuclear Resources*, IPC, London.

2 INCREMENTALISM AND NUCLEAR POWER

1. Partisan mutual adjustment

Lindblom's criticism of synoptic rationality is well known. He argues forcibly that the synoptic ideal is not suited to the realities of practical decision making where intelligence, time and money are all severely limited. One objection is that the synoptic ideal demands unrealistically vast amounts of information, much of which is expensive to obtain. This is certainly so in the case of the breeder reactor. Moreover, Lindblom argues that the synoptic ideal is really empty because it gives no instructions for making choices when anything less than a synoptic view has been achieved. The ideal calls on the decision maker to continue to acquire information until a synoptic view is reached, any choice in the absence of this view being stigmatised as arbitrary. This is clear from the previous chapter where the synoptic ideal poses the policy maker with a dilemma about using data which are still uncertain. The ideal calls for all the uncertainties surrounding the breeder reactor to be resolved before a decision about developing it is made, despite the very long time that is needed for this exercise. If a decision on the breeder is taken using today's best opinions about uncertain technical issues, then the entrenchment of the breeder as it develops may make the use of mistaken opinions very expensive, so the original decision can only be seen as arbitrary by an upholder of synoptic rationality. Either way, the breeder decision will prove to be hugely expensive, in terms of delayed benefits or mistakes that are hard to rectify. This is typical of decisions made according to the rules of synoptic rationality: there is either delay for enough data to be collected, or else there is the risk associated with arbitrary choice.

A further criticism of Lindblom's is that the synoptic ideal requires a very naive view of decision makers. They are portrayed as having a high degree of freedom of choice, at least in the area where it is their business to make decisions, and as having values which are stable at least for periods comparable to the lifetime of the projects they must choose between, and they are supposed actively to seek consistency across their decisions. These features are, however, purely mythological. Policy makers generally have a very small degree of freedom, they cannot move far before coming into contact with other policy makers who are affected by their choices. Somehow or other, a policy maker must manage to live along-side a small army of other decision makers whose interests must be considered and accommodated. Synoptic analysis, which simply tells the policy maker about what best satisfies his own particular values and objectives, is utterly wasted unless other decision makers can be persuaded or cajoled in some way to permit its realisation. It is no good spending a great deal of money to know what is the best option unless the option can be achieved. In general, it cannot be achieved because the interests of other power centres stand in the way.

Many decisions involve very long-term investment: for example, the breeder takes forty to fifty years to recover its development costs. Synoptic rationality therefore needs to assume that the policy maker's values remain constant over this sort of timespan. Not only the values of the policy maker must remain fixed, but also those of the affected groups which have found a place in his synoptic analysis. If the value placed on accidental death or industrial poisoning changes, so that people become more concerned about deaths from breeder accidents or illness from plutonium from the breeder, then the cost-benefit balance is altered in a way which might call for the revision of the breeder decision. This is quite contrary to the philosophy behind the synoptic ideal; the very point of a synoptic view is that the correct decision can be made once and for all, without the need for constant revision and looking over one's shoulder at changing circumstances. This is only possible given a quite unrealistic stability of values.

The synoptic view also pretends that decision makers are concerned to be consistent from one choice to another; indeed if consistency were not a premium virtue there would be little need to indulge in synoptic analysis. It is often remarked, however, that policy makers are hesitant to give a precise and explicit statement of their own values so that this consistency can be ensured. This is usually put down to psychological factors, such as a mild distrust of decision analysts, rather than anything profoundly connected with the problems of making decisions in a world of horrible complexity.

Lindblom presents a much more realistic picture of policy makers in his account of *partisan mutual adjustment*. Lindblom sees each agency as a group of *partisan* decision makers or, for purposes of exposition, a single partisan decision maker. That is, each agency attempts to further its own ends and reach its particular objectives. To achieve its ends any agency must co-operate in some way or another with many other specialised agencies and with partisan groups which are politically active, such as the cabinet, party leaders, groups of MPs, outside pressure groups, the leaders of pressure groups and the ordinary public. The problem, of course, is that the partisan groups with which the first agency must co-operate if it is to achieve its ends may very well have widely different values and objectives. How then can partisans co-operate, how can they mutually adjust in such a way that the interests of all are furthered? Lindblom presents an extensive catalogue of mutual adjustment; he himself sees two classes of adjustment. Where one partisan X simply adapts his or her decisions to those already made by another partisan, Y, Lindblom speaks of *adaptive adjustment*. If X decides without regard to any possible harm that might be done to the interests of Y the adjustment is *parametric*, otherwise it is *calculated* or in the extreme *deferential*, where X actively seeks to avoid injuring Y's interests. The more important second class of adjustments are *manipulated* ones, where X induces some response from Y as a condition of making a decision in a particular way. There are many varieties here:

  (i) *Negotiation* where X and Y induce responses from one another by exchanging information and debating their views.

 (ii) *Bargaining* where X and Y both use conditional threats and promises.

(iii) *Partisan discussion* where X tries to effect a reappraisal of Y's position by giving Y additional information and vice versa.

(iv) *Compensation* where X makes the conditional promise of benefits to Y.

 (v) *Reciprocity* where one or both of the parties calls in an outstanding obligation, or acknowledges a new one.

(vi) *Authoritative prescription* where X uses authority to demand a response from Y.

(vii) *Unconditional manipulation* where X induces a response from Y by unconditionally altering the advantages and/or disadvantages to Y of various responses.

(viii) *Prior decision* where X makes a decision first to induce Y to respond rather than forego the advantages of co-operation with X.

(ix) *Indirect manipulation* where X manipulates some third party in one of the above ways to induce a desired response from Y.

Lindblom is at pains to show two things: that such methods of partisan mutual adjustment actually occur in political decision making, and that these manoeuvres are a more effective way of forming policy than that proposed by synoptic rationality. Decision makers, on this view, are highly constrained by the need to rub along with other decision makers who are seeking to protect their own interests. There is no need to insist that the decision maker's values be constant over decades, because a change in values can be easily accommodated through the process of partisan mutual adjustment. Further, the reticence of decision makers to be precise and explicit about the values they employ is quite understandable. In negotiating with other partisans, it would be a liability to be seen to be committed to some set of values, it would be an obstacle to the usual to-and-fro of negotiation. Thus the need to adjust to the interests of other partisans makes the consistent application of some set of explicit values not only impossible, but a positive obstacle to furthering partisan interest.

A deeper insight into the advantages of Lindblom's partisan mutual adjustment over synoptic rationality requires an explanation of how the partisans are supposed to go about their work. In making decisions, partisans use *disjointed incrementalism* as a way of coping with the problems which we found to vitiate any attempt to employ synoptic rationality. This has the following important features:

(i) *Margin dependent choice*   Decision makers do not attempt a synoptic survey of the problem they have to tackle. Instead, attention is focused on the differences between alternative policies and between these and the status quo. There may be options which alter the status quo greatly, but the magnitude of the innovation makes it impossible to trace out the consequences of the option throughout society and to make intelligent comparisons between these and the consequences of rival options. Such options are, therefore, not considered; analysis is confined to the *marginal* differences between the options and between the options and the status quo. Radical policies are simply not assessed, and options are ranked in order of preference of the incremental differences between them. Such an approach frees the decision maker from the impossibly heavy burden that the synoptic ideal seeks to place on his back, and thus represents an adaptation to his limited intellect, the cost of analysis and the lack of information with which he has to cope.

(ii) *Restricted number of options considered*   The synoptic ideal calls for a listing of all possible options, but considering such a huge and open-ended list is impossibly demanding in terms of intellectual effort, money and information. The problem is eased if only options which are incrementally different are considered, but surveying all of these is still a formidable task. Decision makers therefore restrict themselves by considering only a few of such options.

(iii) *A restricted number of consequences of each option considered*   Having limited the number of options to be analysed, the decision maker is still faced with an impossible task if he attempts to trace out all the consequences of each option. Attention is, therefore, confined to a very few major consequences.

(iv) *Serial analysis and evaluation*   Policy makers never achieve a final synoptic solution to any problem, rather they make a never-ending series of incremental attacks on their problems. For this reason, they do not need to find the 'correct' solution. Their problems are so complex that attempting anything more courageous than a serial attack is simply foolish. This gives the decision maker a great advantage; he can afford to neglect aspects of his problem in the knowledge that he can return to them in the future. A serial attack also prevents the policy maker from discovering the final solution to yesterday's problem, a real possibility given the rate at which many political problems change and the time required for the discovery of a synoptic solution.

(v) *Analysis and evaluation is remedial*   An important feature of real-world decision making is its remedial nature. Policy makers are much less concerned with guiding society to some well defined future state than with finding quick solutions to easily identified social ills. As with the other adaptations, this focus on remedying ills brings about a great simplification in the decision maker's task.

(vi) *Fragmentation of analysis and evaluation*   Policy making is rarely undertaken by one decision maker. Instead, the usual pattern is that different aspects of the same policy problem are considered by a whole number of decision making agencies with different aims and values. This is thoroughly healthy because it means that all important aspects of the problem are likely to receive detailed attention from somebody, even though no one person or agency can attend to the whole problem. The fragmentation of analysis justifies each policy maker paying attention to only part of the problem because parts he neglects are sure to be taken into account by other agencies with interests different from his own. Fragmentation also enables the development of specialist skills by agencies whose task is to consider a particular aspect of many policy problems. If a chosen policy has unexpected bad consequences, one of these specialised agencies is almost certain to identify the unwelcome effect and to consider ways in which it might be eased, given the ameliorative focus of policy formation.

In this, Lindblom suggests, disjointed incrementalism is a strategy that policy makers not only use to overcome the difficulties that stand in the way of following the synoptic ideal, but which they use *successfully*. It is, for Lindblom, a rational way of proceeding, given the kind of problems which policy making faces.

This takes us to a further problem facing synoptic rationality which Lindblom identifies, that of co-ordination. In taking a synoptic view, someone must understand fully a vast body of information usually drawn from a wide range of disciplines. Thus, if whoever has to make a choice about the breeder reactor feels that synoptic rationality should be followed, he had better be a metallurgist, a power engineer, an economist, a inorganic chemist, a nuclear physicist, a nuclear engineer, a works manager, an accountant, a skilled forecaster and so on and so on. The idea that the co-ordination of such a vast quantity of technical information requires a co-ordinator is really absurd, the task is quite beyond the brainpower of even the greatest polymath. Once again, the synoptic ideal is not suited to the realities of limited intelligence. But by what mechanism does partisan mutual adjustment achieve any better co-ordination? Any partisan will pay particular attention to those aspects of a policy problem which affect his own particular interests and motives, a concentration which ensures the development of skills of analysis and negotiation. No single partisan can have a synoptic view of a policy problem, but this is not a ruinous limitation if other partisans pay attention to other aspects of the problem, bringing to bear their own special skills moti- vated by their own particular interests. Lindblom gives a brief example:

A city traffic engineer, for example, might propose the allocation of certain streets to one-way traffic. In so doing, he might be quite unable to predict how many serious bottlenecks in traffic, if any, would develop and where they would arise. Nevertheless, he might confidently make his recommendations, assuming that if any bottlenecks arose, appropriate steps to solve the new problems could be taken at that time—new traffic lights, assignment of a traffic patrolman, or further revision of the one- way plan itself. He might also have quite correctly anticipated certain other consequences, such as business losses from re-routing customer traffic. Some of these he nevertheless ignores as a consideration in his traffic plan. Instead he proposes subsequently to alter parking regulations, ease pedestrian traffic in certain areas, or turn to still some other policy to reduce the business losses ruled irrelevant to this first policy problem. The remedial and serial character of his strategy in effect achieves remedies for emerging problems.

In short, if decision making is remedial and serial, anticipated adverse consequences of any given policy can often be better dealt with if regarded as new and separate problems than if regarded as aspects of an existing problem. And unanticipated adverse consequences can often be better guarded against by waiting for their emergence than by often futile attempts to anticipate every contingency as required in synoptic problem solving. [Lindblom (1965), 150–1]

The whole process of serial remedy is made much more efficient if a large number of partisans are involved in the decision. Each partisan having distinct interests from the others, their multiplicity ensures that no consequences of a decision that may be unwelcome goes unnoticed and unremedied. As well as remedying earlier decisions of another partisan, each partisan may also call the attention of others to problems about decisions not yet made. In Lindblom's words:

The great multiplicity of decision makers in, say, American public policy making can be seen, therefore, as a great strength where problem solving cannot be synoptically accomplished but must be strategically pursued. Multiplicity copes with the inevitability of omission and other errors in complex problem solving. Were there no decision makers with a stake in international trade, we might wonder whether farm policy might not put strains on international trade to which the farm-policy decision makers might themselves be inadequately sensitive; but we know that, if the strains appear, those decision makers who have a stake in international trade will attack them as their own problem. Were it not for decision makers with an interest in parks and recreation, we might wonder whether an urban redevelopment board could be trusted to make decisions on the relocation of commercial houses within a city. [Lindblom (1965), 151]

In this way, the remedy of decisions by other decisions, either in anticipation of the problem or following its detection, serves to co-ordinate decisions made by different partisans. Where one partisan fears that some consequences of a decision proposed or already taken by another partisan will harm his interests, he may employ any of the above listed manoeuvres to try to induce a change of mind on the part of the other partisan, in an attempt to co-ordinate their interests.

An important feature of partisan mutual adjustment is that it tends to lead to agreement about what decisions to make. Through negotiation partisans with widely different aims and objectives can come to agree that some policy is in the interest of all. Partisans are strongly motivated to reach agreement because the extent to which any partisan can achieve his ends is determined by the co-operation he can achieve from other partisans. This ensures that no avenue of co-operation is likely to be overlooked.

In the previous chapter, it was argued that synoptic rationality does not present a realistic framework within which decisions about technology like the breeder reactor can be taken. Having outlined the rival conception of partisan mutual adjustment offered by Lindblom, it may now be asked whether it is appropriate for these peculiarly difficult decisions.

2. Can decisions about nuclear power be made by partisan mutual adjustment?

Nuclear power is a technology that partisan mutual adjustment cannot accommodate. What this means for theories of policy making or nuclear technology will be the subject of the bulk of this book. For now, we need to examine the problems which arise when partisan mutual adjustment tries to fit decision making about nuclear power. The problems have the same origins as before: those features which have been found to make synoptic rationality impossible to accommodate decisions on nuclear power present similar problems for Lindblom's account of policy making. The problems arise because nuclear power represents a *non-incremental* change over existing technologies and because nuclear technology, once adopted, is sure to become *entrenched*.

(a) Non-incremental change

Let us remind ourselves about the differences between electricity generated from thermal nuclear reactors and from coal. Tables 2.1 and 2.2 give a list of some of the major direct and indirect costs which make up the cost of generating electricity in each case. It can be seen that there is no overlap at all. The technologies are so different that of all the major

Table 2.1. Elements of Cost for Nuclear Electricity

Direct
Capital costs of nuclear reactors
Load factors of nuclear reactors
Price of natural uranium
Cost of enrichment of uranium
Cost of uranium fuel fabrication
Cost of reprocessing spent fuel
Cost of nuclear waste management
Cost of decommissioning nuclear reactors

Indirect
Cost of accidents at nuclear plant
Cost of health damage from normal emissions from fuel cycle
Cost of loss of civil liberties to safeguard nuclear material
Cost of proliferation of nuclear weapons
Need to keep nuclear industry in business

Table 2.2. Elements of Cost for Coal Generated Electricity

Direct
Capital costs of coal plant
Load factors of coal plant
Price of coal at pithead
Transport costs of coal
Storage costs of coal

Indirect
Cost of accidents in coal mining and distribution
Cost of health impairment in coal mining
Cost of health impairment due to SO_2 and particulate emissions
Amenity cost of emissions
Cost of acid rain fallout
Need to keep miners in employment

---

cost components of nuclear generation costs, none figure in the cost of coal generation and vice versa. This is enough to show that the shift from coal to nuclear generation is a *non-incremental* technical change. It can be contrasted with an incremental change, like the one considered in the previous chapter where two 250-MW coal generating plants are being considered against a single coal plant of 500 MW. The differences between the two proposals are marginal; the two small plants will cost a little more to build, a sum which can be guessed at fairly accurately with a knowledge of past construction, and will consume a little extra coal, which again will be reckonable. Against this, they will be on stream somewhat earlier than the single large plant, and will contribute to a marginally more efficient system since the chances of both failing together are small. In choosing between the two projects, therefore, analytic attention is brought to bear on just these marginal differences.

Comparing nuclear and coal generating programmes is at quite the other end of the spectrum. The great difference between the technologies means that all the items in Table 2.1 must be considered against all the items in Table 2.2; there can be no focusing of analytic work because there is no margin to focus on. If the difference is granted to be non-incremental, then it may be asked what problems this poses for any attempt to make decisions about nuclear power in the way championed by Lindblom, by a set of partisans each employing disjointed incrementalism. We have already seen how the non-incremental nature of the change from coal to nuclear generation vitiates any attempt to make decisions about nuclear power according to the synoptic ideal. Will it prove as much a problem for Lindblom? It will be best to consider some of the elements of disjointed incrementalism separately.

(i) *Margin dependent choice*   At once, there must be a serious problem because the differences between nuclear power and coal generation are so great that there is simply no margin upon which to bestow our analytical labours. The whole purpose of basing choice upon the marginal differences between options is to avoid being overwhelmed by the kind of demands made by the synoptic ideal for a complete analysis. But this escape seems to be impossible in the present case.

(ii) *Restricted number of consequences of each option considered*   Incremental changes can be handled by partisan mutual adjustment because the number of partisans who are involved is necessarily limited, and the scope for disagreement is also limited because what is at issue are the marginal differences between options. For non-incremental changes, however, the number of partisans who may come to be involved and the scope for argument become open ended. In the present case, the shift from coal to nuclear power is one which may alter many features of society, and so an enormous number of partisans have an interest that may prove to be threatened by the new technology. As widespread an interest in such minute matters as whether two coal stations are built or one large one, or whether a coal station should have cooling towers or not, or whether it is worth paying so much in order to reduce SO_2 emissions from its chimney, would be quite absurd. These are incremental decisions about which only a tiny handful of partisans need be concerned. But the non-incremental change considered here excites partisans of every sort who, quite legitimately, search for ways in which nuclear power might run counter to their own local interest. This effectively prevents the formation of any focus of analysis in the normal way and ensures that the arguments about nuclear power continue. As time goes on, instead of analytic attention focusing on some set of crucial differences between the two technologies, the battlefield becomes increasingly confusing as more and more partisans seek to protect their interests.

Proponents of nuclear technology have always tried to concentrate on the direct costs of nuclear power versus the direct costs of coal generation. This is a difficult enough task, as will become clear later, but its purpose is to focus analytic attention on a part of the difference between the two technologies where the proponents of nuclear power feel they can convince other partisans about the merits of their innovation. In this way, it is hoped that the normal pattern of disjointed incrementalist decision making will be followed and a restricted number of consequences of each option be considered, namely the direct economic cost of the rival methods of generation. But this does not happen. The non-incremental difference between the means of generation makes it impossible for other partisans to accept this as the central problem. Lawyers and civil liberty groups, for example, argue that the dangers of restrictions on civil liberties needed to protect nuclear fuel and wastes are of far greater importance than the economic benefits which might be achieved by nuclear power. Political scientists likewise argue that the dangers of the proliferation of nuclear weapons from the diffusion of reactor technology are more central to the choice between coal and nuclear power than economics. Radiation scientists similarly warn of the radiological hazards of nuclear power, which they see as of much greater importance than bare economics. Residents faced with a nearby nuclear plant express concern about those aspects of the technology which affect them most, i.e. they operate as partisans. Thus they point to the dangers of accidents at nuclear plants and to the dangers of routine radioactive emissions; thus the debate between partisans never manages to focus on economic issues.

In this way, the normal operation of disjointed incrementalism fails, and no consensus emerges about the consequences of each option that ought to be considered.

(iii) *Restricted number of options considered*   The same problem besets attempts to restrict the debate to a limited number of options. Partisans who are afraid of nuclear power, but nevertheless impressed by the economic benefits of the new technology, shift the ground of the debate to other options which they argue to be superior to nuclear power. A favourite here is conservation. It is argued that instead of building nuclear power stations, steps should be taken to encourage the efficient use of electricity so that no extra plant need be constructed. Other partisans argue that additional sources of electricity should be exploited such as wind, wave or solar power. There is therefore failure to agree what options should receive attention, which amounts to a failure of disjointed incrementalism. The non-incremental nature of the nuclear innovation means that a very large number of partisans feel themselves to be affected, and many of these are so distant from the partisans supporting the technology, that consensus on what options to consider is not achievable. To the CEGB, for example, wave power is an option which deserves little attention. It cannot hope, according to them, to meet electricity demand in the coming decades, and were it ever to provide a significant quantity of electricity, it would be very disruptive to the established electricity distribution and generation system with which it must mesh. From the point of view of the CEGB, fear of failing to meet electricity demand and disrupting the supply system are enough to rule out wave power as an option worthy to be considered against nuclear power. All partisans with a similar interest in the efficient provision of electricity will agree, and if these were the only partisans there would be movement towards a consensus about the options to be considered. But for nuclear power, there are many partisans who are totally remote from the electricity industry, and who do not see its problems as their own. For them, the CEGB is being outrageously conservative in not considering wave power, and is perceived as trying to achieve its favoured option by sleight of hand. In this way there is no consensus about what options to take seriously.

(iv) *Fragmentation of decision making*   In the countries where the development of nuclear power began, it was recognised very early that its special features called for special organisations for its promotion and control. These had different structures and authority, but they all played an important part in the development of nuclear power in the country concerned. The great complexity of nuclear technology was itself sufficient to ensure that the partisan specially charged with its development would have a powerful control over the technology. In all countries, this led to a centralisation of power in nuclear matters, which ran counter to Lindblom's aim of fragmented decision making, as will become clearer in the historical section to follow.

In Britain, the inevitable concentration of power in the hands of the partisan equipped with the technical information concerning the new technology was re-inforced by the strong centralising tendencies of British administration; the UKAEA was set up as a monopoly provider of technical advice on nuclear power, and of nuclear technology up to and including the prototype of whatever reactor designs it chose to develop. Its operation is supposed to be under the scrutiny of a government ministry, currently the Department of Energy. In practice, however, the scrutiny has proved to be purely nominal. Efficient scrutiny obviously calls for a scrutineer with experience in nuclear technology, so that the UKAEA's opinions, for example on such matters as reactor design, could be challenged. But such people have been found only in the UKAEA itself, and in the commercial nuclear industry. No government department which has been charged with overseeing the activities of the UKAEA has been able to build up a group of civil servants sufficiently versed in nuclear technology to be able to do their job properly and to provide an independent voice for ministerial policy makers and an independent measure of the UKAEA's work. Power has therefore centred very much on the UKAEA in the making of decisions about nuclear technology in Britain. This is far from the end of the story, of course, which would involve secrecy, the capture of the electricity generating industry by the UKAEA, the transfer of personnel between these two groups, and much more. But this is not the place for such a lengthy digression. For present purposes, it is enough to note that the UKAEA has held great power in the directing of nuclear policy, to such an extent as to breach Lindblom's demand for many partisans to be involved. The British example could be matched by others from the United States and France. Something will be said about this in the historical discussion.

Nuclear power therefore presents a paradox. Its novelty and non-incremental nature mean that many partisans of all varieties feel affected by it and feel moved to defend their particular partisan interest from the threat of the new technology. On the other hand, these same features lead to a centralisation of power within special partisans who have the technical expertise needed to exploit the technology and who form the natural source of advice to policy makers. The debate about the merits of nuclear energy is therefore conducted between, on one side, many partisans equipped with little technical experience and having, at best, second-hand knowledge, and, on the other side, a few very powerful partisans equipped with great knowledge and having the ear of government policy makers. This is contrary to the efficient working of Lindblom's disjointed incrementalism, where power should be widely distributed among the interested partisans.

In the above section, I have not tried to give a detailed analysis of the nuclear debate which has been raging in so many countries for so many years, for this would require a book or two by itself. I have instead tried to show, with a few illustrations, what problems the non-incremental nature of nuclear power makes for Lindblom's partisan mutual adjustment.

(b) *Entrenchment*

The way in which the breeder reactor would become entrenched has been examined briefly in the previous chapter. The same can be expected of nuclear power in general. In its infancy, a technology's development may be channelled in all kinds of ways, but not enough is known about the unwanted social effects it might have for it to be directed in such a way as to avoid their occurrence. By the time it is known that the technology has unwanted social effects it is often the case that the technology has become so much part of the whole social and economic fabric that little can be done to control it in any fundamental and radical way. Whatever controls are then possible will be *ad hoc* and superficial. Any attempt at a more radical control will be very expensive, slow to operate and full of disruption and dissent because, at this stage in its development, a radical change in the technology will involve changing all the surrounding technologies and social institutions which have adjusted to its existence. At the time when changing a technology's path is easy, not enough is known about its social effects for those which are unwanted to be avoided. When the technology has become mature enough for these effects to be manifested, then little can be done to control it. This is the *dilemma of control* (Collingridge (1980)).

This is sure to happen to nuclear power if it is ever responsible for a large fraction of electricity production. Devices which use electricity, such as lights and motors, cannot be changed to other fuels, wholly new equipment must be installed for this, such as gas lamps or internal combustion engines. If nuclear power is found to have some unwanted social effect, such as causing many deaths from radioactive emissions or plant explosions, it would therefore be very expensive for nuclear stations to be closed down in order to avoid the problem. Electricity generation could not, at least in the short-medium term, be switched to other fuels such as coal, since there will be neither the mines to produce the coal nor the boilers to burn it. The loss of nuclear capacity would mean a drastic impairment of the economy. Electricity-using equipment could not be substituted for many years and so there is little prospect of reducing electricity demand. But many years and huge expense would be involved in replacing nuclear stations with non-nuclear ones. Thus only marginal, *ad hoc* controls would exist for the problem, such as a slight shift to other fuels, or add-on pollution or safety devices. The root and branch control of closing down nuclear plant would be ruinously expensive. This has problems for Lindblom's views, which as before can best be illustrated through the obstacle entrenchment poses to the efficient working of disjointed incrementalism.

(i) *Evaluation and analysis is remedial* The first problem for disjointed incrementalism concerns the remedial focus of policy making and the serial nature of analysis and evaluation. The remedial focus of policy requires that as much room as possible be left for controlling whatever problems a particular policy might lead to. In this way there is a good chance of being able to find a control which is satisfactory to the other partisans who are interested. The reduction of control options makes remedy difficult and so runs counter to the spirit of partisan mutual adjustment. But this is, of course, just what happens in the case of nuclear power. If some partisans want to control a consequence of nuclear power which has come to light at a time when it is supplying a significant fraction of electricity demand, then there is little room for control, since only *ad hoc* remedies will now be feasible.

(ii) *Serial analysis and evaluation* Entrenchment also poses a problem for the demand made by disjointed incrementalism that analysis be serial and continuing. There is no point in analysis unless its results can be used to guide policy making. If little or nothing can be done to resolve a problem, there is no point in spending money to obtain a more detailed knowledge of the difficulty. But this is just what happens when a technology becomes entrenched; there are so few ways in which its unwanted social effects might be controlled that there is little point in detailed and serial analysis of the technology.

In the discussion to follow, it will be found that one of the properties which encourages the entrenchment of nuclear power is its capital intensity. Suppose that a nuclear reactor is built on the basis of forecasts which show it to be a cheaper source of electricity than any other option. If the forecasts prove to be inaccurate, and, for example, coal is found to be a cheaper fuel than nuclear power, then there is little room to act on this sad discovery even if it occurs before the reactor is built. Let the capital costs of the nuclear plant be C_N, fuel costs for lifetime generation F_N, and the capital cost sunk in construction by, say, year six, C_N^6. Economics points to abandoning the reactor if the capital costs of an equivalent coal station plus the cost of coal for equivalent generation is less than $(C_N - C_N^6) + F_N$. This is very unlikely to happen; by year six so much capital is likely to have been sunk in construction that it will be uneconomic to stop building it and switch to a coal station. Once investment in a nuclear plant has reached a certain level, it is expensive to do anything but continue its building. Given this, there is little point in agonising about generating costs from coal; whatever is found, nothing can be done. Serial analysis in this situation is simply an empty decoration.

To conclude this chapter; there is a clash between Lindblom's partisan mutual adjustment and nuclear power. Nuclear technology has features which make it impossible to make decisions about it using the methods favoured by Lindblom. In which way will the clash be resolved?

References

Braybrook, D. and C. Lindblom (1963), *A Strategy for Decision*, Collier-Macmillan, London.

Collingridge, D. (1980), *The Social Control of Technology*, Frances Pinter, London.

Lindblom, C. (1965), *The Intelligence of Democracy*, Collier-Macmillan, London.

Lindblom, C. (1968), *The Decision Making Process*, Prentice Hall, Englewood Cliffs.

Lindblom, C. (1979), 'Still Muddling, Not Yet Through', *Public Administration Review,* **39**, 517-26.

3 A TEST FOR PARTISAN MUTUAL ADJUSTMENT

The previous two chapters have identified a clash between two theories of decision making, synoptic rationality and partisan mutual adjustment, and choices about nuclear power. It appears impossible to make decisions about the technology of nuclear power in the way enjoined by either theory of decision making. The task of this chapter is to gain a better understanding of this clash and the various ways in which it might be resolved. The first move towards this improved understanding is to remind ourselves about the nature of the two theories of decision making which have been considered. Each one is both normative and descriptive; it states how decisions *ought* to be made by agents in pursuit of their own interests, and it also describes how choice *is* made. These two aspects are intimately connected in a way which may be illuminated by considering Popper's theory of scientific method. Popper's theory of science lays down methodological rules that ought to be followed in the pursuit of truth. Now science appears to be extremely successful in furthering our understanding of the universe and in providing answers to the more mundane practical questions which we pose to it. If a firm finds that a batch of the glue it sells fails to stick whatever it is supposed to stick, it may wish to discover why and what can be done to prevent its recurrence, in which case the problem is passed to the laboratory where answers are sought from those initiated into the mysteries of chemistry, physics and so on. The answers obtained in such a way tend to be superior to those that might be offered by sheer guesswork, navel gazing, consulting oracles and so on, and there is a hard-nosed test for their superiority, the stickiness of future batches of the firm's glue. So science seems to be successful in ways little and big a thousand times a day. This needs to be explained in some way—why is science more successful at solving a particular class of problems than all the other ways which might be taken?

One explanation is that the secret of success is that science actually follows the methodological rules laid down by Popper, and so represents a rational pursuit of truth. Scientists are successful in tackling problems because the methods they use are ones appropriate to the pursuit of truth. This explanation of the success of science must be strongly favoured by Popperians. Any other explanation would have to view the success of science as an illusion or as a huge accident. If science does not use Popperian methodology, it cannot be seen as a rational pursuit of truth, and any success it may have must either be merely apparent, perhaps because other systems of problem solving have not been explored in sufficient detail, or purely coincidental. To avoid such embarrassment, Popperian philosophers have attempted to see the actual practice of science as a reflection of Popperian methodology in action.

The same is the case for decision theories. Synoptic rationality, for example, prescribes rules that ought to be followed by a rational agent who seeks to further his own interests. But it also claims that something like these rules are actually used in practical decision making, either literally, or as an ideal towards which real decision makers aim. In this way, the theory can explain why decision making tends to be successful. It undoubtedly seems better to pursue one's aims by making active decisions rather than waiting passively in the hope that fate will deliver what is wanted. This must be believed, or else we would never bother to make decisions, becoming passive objects tossed haphazardly by whatever happens around us. Synoptic theorists must claim that thoughtful and informed decision making is successful because the rules which ought to be followed by agents in furthering their own interests are actually followed. Decision makers must be seen, that is, as employing something like the normative rules which are entailed by synoptic rationality. In this way, the normative aspect of the synoptic theory is central; whatever success it might have in describing actual decision practices depends upon it first being clear about what practices ought to be followed. Partisan mutual adjustment has the same double aspect as synoptic rationality. At root, it is a normative prescription about how choice ought to be made by a set of agents seeking their own interests. The theory can then explain why making decisions seems a reasonably good way of getting what you want, by showing how decisions in the real world are made in something like the way the theory holds they ought to be made. As before, this makes the normative aspect of partisan mutual adjustment primary. The success of making decisions in the real world can only be explained once it is clear how decisions ought to be made.

What does this tell us about the problem posed by decisions on nuclear power? If the two decision theories were purely descriptive, the discovery of a class of decisions, those concerning nuclear power, which cannot be made in the way described would amount to a falsification of each of the theories. If synoptic rationality claimed that, as a matter of fact, all decisions are taken in the way it describes, then decisions on nuclear power would falsify this claim. If partisan mutual adjustment claimed that, as a matter of fact, decision makers always behave in the way partisans are described as behaving, then the discovery of a set of decisions for which this could not be the case would falsify the theory. Whether these falsifications would be counted as sufficiently serious to force the abandonment of the decision theories, or whether the problem would be resolvable in some less traumatic way, is a question we can happily avoid. But this would be to forget the primarily normative aspect of the two theories. The problem posed by nuclear power is far more interesting, compelling and fruitful than any simple falsification of universal descriptions about human choices. Both theories state how decisions ought to be made, but choices about nuclear power present a case where decisions cannot be made in the way they ought to be made. This is far more interesting than the discovery of choices which cannot be made in the same way as other decisions are. Consider, for example, the response of someone charged with steering the development of nuclear technology. To be told that the decisions that have to be made cannot be made in the way most other decisions are made is perhaps a compliment about the special status and importance of the technology. But to be told that decisions about nuclear power cannot be made as they ought to be is to be presented with the greatest sort of worry: all one's efforts to achieve good decision making, all the labours of experiment and data gathering, all the traumas of debate and argument, and all the reports written and read amount to nothing more than a brave attempt to do the impossible because despite them all decisions about nuclear power must be flawed. This is a challenge indeed.

There are two ways of responding to the discovery of some

set of choices which cannot be made in the way prescribed by a general theory about how decisions ought to be made. If the decisions causing the problem seem perfectly ordinary ones, then the theory will be judged to be inadequate for it fails to lay down rules for the making of these straightforward choices. The other possibility is to keep the theory and show that there is something abnormal about the class of decisions to which it fails to apply. Lindblom's criticisms of synoptic rationality involve a response of the first kind. He argues, it will be remembered, that the rules laid down by synoptic rationality for successful decision making are so severe that any attempt to apply them beyond some very simple games inevitably fail. Seeing the rules as ideals towards which decision makers ought to aim does nothing to resolve the problem because synoptic rationality provides no prescriptions which can be followed in cases which are less than ideal. The inability of synoptic rationality to cope with decisions about nuclear power is therefore not a special problem. It is, rather, an extreme case of a very general problem. The shortcomings of synoptic rationality are particularly clearly illustrated by its inability to cope with decisions about nuclear power, but less obvious failures abound. In the rest of this work I therefore propose to concentrate upon the rival view of partisan mutual adjustment, synoptic rationality being rejected because of its extremely limited scope, and not simply for its inability to provide a framework for decision making about nuclear power.

The question therefore becomes one of finding a suitable response to the inability of partisan mutual adjustment to provide rules for choices concerning nuclear power. Parallel argument would seem to indicate the rejection of partisan mutual adjustment, until it is remembered that synoptic rationality was rejected not just because of the problems it ran into over nuclear power, but because these problems were generic, arising, in more or less obvious ways, whenever the theory is applied to real-world decisions more complex than simple games. This is not the case, however, for the rival view. Partisan mutual adjustment can be applied to many choices where synoptic rationality fails; for a recent discussion see Collingridge and Douglas (1982). The problems posed by nuclear power cannot therefore be seen as generic ones, arising whenever attempts are made to apply the rules of partisan adjustment. This leaves open the second response considered above. Partisan mutual adjustment could be retained if decisions on nuclear power can be shown to be aberrant in some way as to make them inherently difficult to manage. Showing this would be a notable success for partisan mutual adjustment. The theory has singled out a class of decisions which cannot be made according to its prescriptions. If it is then shown, quite independently of partisan mutual adjustment, that these decisions are peculiarly risky and open to uncontrollably high costs, then this marks a great success for the theory. Success will be enhanced if the very same features of nuclear technology which make it peculiarly risky are those very features which make partisan mutual adjustment inapplicable to nuclear power. If this could be shown, then the inability to accommodate nuclear power decisions would have to be seen, not as a shortcoming for partisan mutual adjustment, but as a considerable success. The decisions to which the theory fails to apply turn out to be not ordinary choices at all, but ones which are extremely risky and prone to expensive error.

Any theory about how decisions ought to be made is bound to be inapplicable to some decisions. Partisan mutual adjustment, for example, cannot be applied to decisions which are taken completely at random. This is hardly troublesome because we know these decisions to be reckless quite independently of anything partisan mutual adjustment tells us about them. The case of nuclear power is more exciting because until now decisions about this technology have been seen as perfectly ordinary and comparable to any other investment in new technologies. If this view is retained, then there is a severe problem for partisan mutual adjustment because it claims to provide rules for such choices. If, on the other hand, arguments which are independent of partisan mutual adjustment can show that, contrary to current perceptions, nuclear technology involves risks which are quite out of comparison with those involved in other technological ventures, then the theory may be retained, and counted as scoring a notable success.

The test which I propose may be formalised in the following way:

1 Partisan mutual adjustment → If an agent wishes to further his own interests, then he should make all decisions by rules $R_1 \ldots R_N$

2 Decisions to invest in items with features F cannot be made by rules $R_1 \ldots R_N$

3 Nuclear power has features F

1. 2. 3 entails 4

4 If decisions to invest in nuclear power further an agent's own interests, then not (partisan mutual adjustment)

From statement 3 we know that nuclear power has features F, in which case, by 2, decisions about investing in it cannot be made by $R_1 \ldots R_N$. If such investment decisions further one's own interests, then this cannot be because they are made by the rules $R_1 \ldots R_N$, in contradiction to partisan mutual adjustment by 1. If nuclear power is regarded as any other technology, then it will certainly be possible for decisions about investing in it to serve the chooser's interests, in which case partisan mutual adjustment will have to be given up. Putting it formally:

5 Deciding to invest in nuclear power furthers an agent's own interests

4. 5 → Not (partisan mutual adjustment)

Thus, 1. 2. 3. 5 amount to a falsification of partisan mutual adjustment. If a falsification is attempted, and fails, then the theory being tested is corroborated and must be accounted a success. Corroboration in this case would result if it could be argued that investment in nuclear power involves such uncalculable risks that it does not serve anyone's interests, i.e. if the contrary of 5 can be shown, i.e.

6 Deciding to invest in nuclear power does not further an agent's own interests

If 6 is accepted, 1. 2. 3 poses no problem for partisan mutual adjustment. I hope to strengthen the test further by arguing for 6 indirectly through the general assertion

7 Deciding to invest in items with features F does not further an agent's own interests

so that

$$3.\ 7 \rightarrow 6$$

In this way it is shown that the very features, F, which make it impossible for partisan mutual adjustment to accommodate decisions about nuclear power make such decisions so peculiarly risky that they cannot serve anyone's interests.

This is the logical form of the test proposed here for partisan mutual adjustment, but what is going to be F? In Collingridge (1982) I proposed a theory of social choice which I called *critical decision theory*. It is an attempt to construct a Popperian theory of decision making which is sceptical in the sense of denying that any choice can be justified by the chooser. The question then becomes 'if any decision may prove to be mistaken, how can error be detected quickly and remedied swiftly and easily?'. Decisions are to be exposed as erroneous by the discovery of new, or previously overlooked facts which, once recognised by the chooser are sufficient for him to alter his original preferences. I tried to show that critical decision theory could provide a philosophical base for Lindblom's partisan mutual adjustment with which it shares a number of features. Critical decision theory entails partisan mutual adjustment with two important amendments. For Lindblom, partisans have to adjust their choices to the decisions of other partisans, for which purpose they have a very large number of strategies, which were outlined in the previous chapter. But Lindblom does not discriminate between these strategies, they are all useful if they lead to the consensus required of the partisans. From the perspective of critical decision theory, however, there is a profound difference between those strategies which involve the transfer and discovery of information, thereby promoting better decision making by exposing previous errors, and those which involve the purely accidental power distribution between partisans, such as the strategies of authoritative prescription or compensation which reflect power and money more than good argument. The former strategies are the only ones acceptable to critical decision theory, although there may have to be a place for the second sort in real-world decision making.

For present purposes the second amendment is more important. It concerns *flexibility*. Lindblom occasionally refers to

the need for partisans to retain flexibility if they are to be
able to adjust to one another, but this receives little explicit
discussion, perhaps because Lindblom underestimates the
problems of achieving flexibility (Collingridge (1980)). Critical
decision making places central importance on the maintenance
of flexibility in the sense of the ability to change a decision
if new information comes to light which reveals it to be mistaken. If any decision is open to error, it follows that all
decisions should be open to reversal if error is discovered after
they are made. This, I want to argue, is the key to understanding the problems generated by nuclear power. Nuclear power,
in short, is a *highly inflexible* technology, and it is this feature
which is to be our *F*. It is because of its inflexibility that
decisions about nuclear power cannot be accommodated
within the prescriptions of partisan mutual adjustment, and
this same inflexibility makes decisions about nuclear technology
so peculiarly unmanageable and risky.

Part II of this work, which follows immediately, will therefore argue, quite independently of partisan mutual adjustment,
that nuclear power is a very inflexible technology, so much so
that making decisions about it are best avoided by those seeking
to serve their own interests. It will be argued that: making
decisions about items which are highly inflexible does not
further an agent's own interests; and nuclear power is highly
inflexible.

In this way 6 above, deciding to invest in a nuclear power
does not further an agent's own interests, will be defended
with its consequent corroboration of partisan mutual adjustment. This will require an extensive discussion of the history
of civil nuclear power, which will cover the next four chapters.

In this chapter I have so far refrained from saying anything
about the various views of policy making which claim to fall
between synoptic rationality and partisan mutual adjustment,
and this must now be remedied. Etzioni (1967, 1968) accepts
Lindblom's criticism of the synoptic ideal, but rejects disjointed incrementalism on the grounds that it gives too great
a weight to the powerful in policy making, that many decisions
are not incremental and that it encourages bureaucratic inertia.
Etzioni seeks a compromise position by adding two features to
disjointed incrementalism. First of all, policy making bodies
are to concern themselves not solely with everyday operational

details, but are also to devote some of their energy to scanning the environment, both near and distant, for issues which might demand attention. Secondly, Etzioni distinguishes between fundamental and incremental decisions. The former set the direction for a series of incremental decisions, whose value and function is determined by the fundamental decision. For fundamental decisions, incrementalism is inadequate. The policy maker ought to consider all the main alternatives for such choices, trying to eliminate those options which reveal crippling objections. Not all options need to be considered in the same depth, for as soon as such an objection is found, the option requires no further attention. The implementation of a fundamental decision is to be flexible and serial, and scanning should search for problems which its implementation is causing.

Closer inspection shows, however, that Etzioni's efforts do not really provide a satisfactory theory of policy making. Scanning adds nothing to disjointed incrementalism. As Lindblom recognises, scanning, in the sense of searching the environment for features which might imply change in an organisation, cannot be conducted efficiently by the organisation itself. Organisations have a vested interest in what happens in their environment and so tend to perceive events in a way which does not embarrass them. Correction is therefore likely to come from other agencies with other interests, who can argue that a particular change has occurred in the environment, and can pressure the first organisation to respond to this change. Efficient scanning therefore calls for policy making to be fragmented in just the way Lindblom advocates. It is also promoted if policy has a remedial focus, organisations searching the environment for features which they seek to avoid, and is serial, the search being continuous. But again, these are features of disjointed incrementalism. These features of disjointed incrementalism therefore ensure that this kind of policy making involves efficient scanning in Etzioni's sense. No additional prescriptions are required.

The second novelty suggested by Etzioni, the distinction between fundamental and incremental, collapses in the same way. Etzioni's error is in assuming that fundamental decisions cannot and should not be incremental. Decisions are nested in a well-known way. A firm's decision to lower the level of

lead in the petrol it sells may follow from a decision to make
higher levels illegal, and this in turn may follow from decisions
to improve the environment generally and to set up the necessary apparatus. We may therefore think of decisions which are
'relatively' fundamental. It does not follow, however, that these
are not nor should not be incremental. The decision to reduce
the maximum level of lead permitted in petrol seems to have
been taken quite incrementally, and yet it is 'fundamental' to
the many decisions of petrol suppliers to lower their lead levels.
The existence of relatively fundamental decisions does not
mean that they need to be separated out for special attention
in a way which makes them non-incremental. It may be, of
course, that relatively fundamental decisions require more
thought and analysis, but this everyday observation can hardly
be inconsistent with disjointed incrementalism. All that Etzioni
can realistically claim now is that *some* relatively fundamental
decisions should be made in a non-incremental way. Without
a criterion for identifying which decisions merit this honour,
the claim is a very weak one, and as such of little interest.

   The final criticism of mixed scanning concerns the way in
which the supposed non-incremental decisions should be
taken. As Camhis (1979) observes, Etzioni assumes that policy
makers have predictive powers, which they do not possess in
reality. In eliminating options in making a fundamental decision
less needs to be known about each option than that called for
by synoptic rationality, which seeks to find the best option,
but nevertheless a great deal needs to be known if an option
is to be eliminated, very often more than is reasonable to
demand (Collingridge and Douglas (1982)). For fundamental
decisions all 'main alternatives' need to be considered, Etzioni
urging this as a counter to the inertia of organisations he
supposes to be encouraged by disjointed incrementalism. But,
as we have seen in chapter 1, options involving more than an
incremental change soon begin to require huge quantities of
technical information. Etzioni seems oblivious of these problems and the embarrassment they would cause to policy making
if his proposals were adopted. Similar criticisms might be
levelled at Gershuny (1978); see too Smith and May (1980).

References

Camhis, M. (1979), *Planning Theory and Philosophy*, Tavistock, London.
Collingridge, D. (1980), *The Social Control of Technology*, Frances Pinter, London.
Collingridge, D. (1982), *Critical Decision Making*, Frances Pinter, London.
Collingridge, D. and J. Douglas (1982), 'Three Models of Policy Making—Expert Advice in the Control of Environmental Lead', Research Paper, Technology Policy Unit, University of Aston, Birmingham.
Etzioni, A. (1967), 'Mixed Scanning, A 3rd Approach to Decision Making', *Public Administration Review,* **27**, reprinted in A. Faludi (ed.) (1973), *A Reader in Planning Theory*, Free Press, Glencoe, pp. 217-29.
Etzioni, A. (1968), *The Active Society: A Theory of Societal and Political Process*, Free Press, Glencoe.
Gershuny, J. (1978), 'Policy Making Rationality—a Reformulation', *Policy Sciences,* **9**, 295-316.
Smith, G. and D. May (1980), 'The Artificial Debate Between Rationalist and Incrementalist Models of Decision Making', *Policy and Politics,* **8**, 147-61.

PART II

THE TROUBLED HISTORY OF NUCLEAR POWER

4 A HYPOTHETICAL REACTOR PROGRAMME

This part of the book concerns the history of civil nuclear power. What I hope to show, without any reference to partisan mutual adjustment, is that nuclear power is a highly inflexible technology and that this has made it a very troublesome innovation. The troubled history of nuclear power in Britain and the United States has been much discussed. My excuse for adding to this literature is that most of the existing commentary focuses on what is really a secondary issue. Mistakes were made in the development of nuclear power, and much effort has gone into unearthing just what errors were made and trying to explain them. This is interesting enough but it does not get to the heart of the issue. Mistakes are bound to occur, every day and in every way, in an innovation as complex as nuclear power. What really separates nuclear power from other innovations is not the *existence* of misjudgement, wishful thinking, bad management, mistaken forecasts and plain bad luck, but the extraordinarily high *cost* which these errors incurred. This part of the book seeks to explain this peculiar feature of nuclear power, and in so doing points to lessons about the development of breeder reactors. Collingridge (1980) has previously shown how technologies which are inflexible can impose heavy costs if mistakes are made in their adoption, development and control. Here, I hope to show that it is the inflexibility of civilian nuclear technology which is at the heart of its troubled and costly history. Reactors have become larger, have taken longer and longer to complete, have become increasingly capital intensive and ever more dependent upon infrastructure, such as fuel reprocessing and uranium enrichment. These features combined to ensure that whatever mistakes were made in developing civilian nuclear power, they were slow to be discovered, slow to be remedied and were very expensive. In other words, the technology itself is highly inflexible. Decisions about new technologies such as nuclear power are necessarily made under very great uncertainty, or *ignorance* as I prefer to call it, Collingridge (1980). For such decisions flexibility is an essential consideration; mistakes, if they are made, should be discovered and discovered early, and the ability to revise the original decision in the light of dis-covered errors should also be retained.

The central idea behind this concept of flexibility has been explored in Collingridge (1980, 1982), and is a simple one. Con-sider a system that interacts with the environment surrounding it in ways that confer various benefits and impose various costs on the system's controller. The controller receives signals that tell him how the system is behaving and he can alter its behaviour by adjusting one or more of a number of decision variables of the system, each adjustment taking time to become effective. We may think of the controller as steering the system through the environment by means of the system's decision variables. The payoff over time is a function of the interaction of the system and environment, but in the cases which are of interest to us this cannot be predicted because the controller has to make decisions about the system's decision variables under ignorance. If the controller has to choose which of two systems to steer he cannot, therefore, base his choice on knowledge of the payoffs he will receive from each system. His choice can only be based on his knowledge of the system, not on its inter-action with the environment. One system may be easier to steer than the other; in other words errors in one system's behaviour, its delivery of costs instead of benefits, may be more easily corrected than errors in the other. This will be so, for example, when the systems are identical except that one delivers more signals about its present state to the controller, or where one has more decision variables or where one has the same number of decision variables as the other but the system responds more quickly to changes in them. The controller should obviously choose the system that is more easily controlled, whose errors can be identified and remedied more easily, so that the pay-off over time is less sensitive to error—in other words, the most flexible system. Flexibility is thus to be judged from the system itself, no information about how the system will actually interact with its environment being needed.

This is the thinking behind the measures proposed here. A decision is easy to correct, or highly flexible, when—if it is mistaken—the mistake can be discovered quickly and cheaply and when the mistake imposes only small costs which can be eliminated at little expense.

Elements relevant to flexibility are:

(i) *Monitoring*　The search for facts capable of showing a decision once made to be erroneous may be called *monitoring* the decision. The minimum period from the decision to the discovery of error may be called the *monitor's response time*. Options that have a low monitor's response time should be favoured. If a decision is mistaken, it pays to discover the mistake quickly.

(ii) *The cost of error*　When a wrong decision has been made costs, though not necessarily monetary ones, are imposed; indeed it is these costs which constitute the error. The cost of an error is the decision's *error cost*, which is generally a function of time. If no corrective action is taken, we may speak of *uncontrolled error cost*, remedial action reducing this to a *controlled error cost*. Options with a low controlled error cost should be favoured if a decision has to be made in a state of ignorance. If the chosen option proves to be mistaken the mistake need not involve the bearing of great costs.

(iii) *Time for correction*　A remedy for a discovered error generally takes time to operate fully. The period from the remedial action to the elimination of error cost may be termed the *corrective response time*. The sum of this and the monitor's response time is the *gross response time*. Options with a low corrective response time should be favoured in making decisions under ignorance. There are three reasons for this. Remedying a mistaken decision quickly means that error costs are eliminated quickly and benefits from an improved decision are obtained early, remembering that early benefits are to be valued more highly than postponed benefits. Secondly, ignorance often extends to the effectiveness of the imposed remedy. When this is so, a remedy with a low corrective response time may be discovered to be ineffective more quickly than a remedy with a high corrective response time. This means that another, and perhaps more effective remedy, can be substituted more quickly. The sooner the substitution, the lower the error cost. Thirdly, a low corrective response time leaves the decision maker with more options once the discovered mistake has been remedied.

(iv) *The cost of remedy*   The cost of applying a remedy for a mistaken decision may be called the *control cost*. Low control cost is obviously a desirable feature of decisions made under ignorance, but, in addition, where the effectiveness of the remedy is not known favour should be given to remedies having a high variable to fixed cost ratio. If the remedy is found to be ineffective, then all the fixed costs are generally lost, but some of the variable costs will be avoidable once another remedy is substituted.

This, briefly, is the concept of flexibility that is to be used in explaining the cost of the errors occurring in various nuclear power programmes, as outlined in Collingridge (1983). But first it may be useful to consider the conclusions of other commentators. The chief lessons drawn by earlier commentators concern principally the institutional arrangements for nuclear power. Williams (1980a and 1980b) talks of the poor accountability of public bodies in Britain, particularly the CEGB and the UKAEA, and sees the need to improve this feature of the country's administration as one of the chief lessons of the British nuclear power story, a sentiment shared by Henderson (1977). Franks (1983), Patterson (1977) and Rush *et al.* (1977) are particularly concerned by the dominating position of the UKAEA as supplier of expert opinion and monopoly supplier of prototypes, and the deep-seated secrecy surrounding much of its affairs. To counter this, Sweet (1978) has suggested an independent body to scrutinise estimates of nuclear costs, part of a more general plea for greater access to information. Wonder (1976) shows how the attention of bodies who might have been able to scrutinise in this way was constantly diverted from the real issues of reactor choice and directed to side issues of industrial organisation. Hinton (1980), on the other hand, argues that the organisations making choices about nuclear development in Britain were not centralised enough.

Bupp and Derian (1981) point to the overconfidence of suppliers, USAEC, JCAE, and utilities in the level of development of nuclear power in America. Development of reactors should have been much slower, so that decisions could have been made on the basis of operating experience and not the rosy forecasts of the new technology's proponents engendered by competition between suppliers. For Burn (1967, 1978), the better performance of United States suppliers is to be explained by their investment in many prototypes, in contrast to the narrow-front philosophy of the UKAEA, a reflection of the very different attitudes the two countries have to commercial competition.

All of these claims concern what mistakes were made, why they were made and what might be done to avoid similar errors in the future. But this is only a start to understanding the troubled history of nuclear power. Mistakes were bound to happen in an enterprise of this magnitude. What is impressive about nuclear power is not the existence of these errors, but the quite unprecedented cost which their making entailed. To understand this feature of the innovation, we need to go beyond the organisational surroundings to the technology itself. In this way we might be led to much more profound lessons than have been drawn before.

In detail, I hope to show the following:

(1) Misjudged investment in commercial nuclear plant occurred in Britain with the Magnox and AGR programmes, and in the United States with LWRs.
(2) The cost of misjudgement in each case proved to be very large and to have four components: (a) misinvestment in nuclear plant; (b) the inability to rescue the situation by shifting from nuclear to other fuels; (c) restrictions on future nuclear development; and (d) the need to maintain nuclear infrastructure.
(3) The great cost of these misjudgements is due to four features of the technology itself; (a) the long lead time of plant; (b) the large unit size of plants; (c) the plants' high capital costs and capital intensity and (d) the requirements on infrastructure made by the plants.
(4) The apparent success of the French nuclear programme masks the very great risks which were taken and which still threaten the programme, the magnitude of the risk being determined by the same four features of the technology noted in (3) above.

This historical discussion will constitute the greater portion of Part II, but before discussing the messy and confusing details of real history, it will be useful to sharpen the eye of analysis through discussion of a completely hypothetical reactor programme. Armed with whatever insights this provides, we may then tackle history in a more systematic way.

Consider a hypothetical reactor which, like real ones, is capital intensive; has a long construction or lead time; is built in large units in an attempt to gain economies of scale; and depends for its operation on a nuclear infrastructure which ranges from uranium mining, processing, enrichment and fabrication to reprocessing and waste disposal and includes all the trained specialists and plant designers and builders working specifically for the nuclear industry. We may consider the problems of introducing this reactor to fully commercial operation from its prototypes. We may assume that competition comes from coal-burning generating plant, which will be taken to provide the present cheapest base load plant. For present purposes a mistaken investment in such hypothetical nuclear plant is its construction when electricity could have been produced more cheaply by a coal burning station (or stations) of the same capacity ordered at the same time. Restricting the discussion to economics in this will not reduce the force of the arguments to be developed, but will enable us to sidestep discussion of the more contentious problems posed by the technology, such as the proliferation of nuclear weapons or the loss of civil liberties needed to safeguard the transport and storage of plutonium. There are two questions to be asked: how likely is it that such mistakes will be made and, if they are made, what costs will be incurred?

1. The likelihood of error

Taking the first question, it may be observed that the choice between nuclear and coal programmes (or a single nuclear and coal plant) requires the forecasting of a great many factors, the most important of which are shown in Table 4.1. Some of the components are under direct human control and others are not. Thus, engineers will try to reduce the capital costs of successive nuclear reactors as they learn how to construct them, and their growing experience can be expected to bring about similar improvements in load factors. A rival gang of engineers will try to do the same for coal plant, thinking of ways to reduce capital costs and to get more electricity from each ton of coal which is burned and to increase the load

Table 4.1. Important Forecasts in Choosing Between
Thermal Reactor and Coal Plant Programmes

Capital costs of the nuclear plants
The load factors that will be achieved by the nuclear plants
Uranium prices
Capital costs of coal plants
Coal prices
Electricity demand

---

factors of coal plants. Forecasts of what these engineers will achieve in the future therefore need to be made.

As we shall see in following chapters, a common assumption here was that a steady reduction in the capital costs of nuclear plant would be achieved, with a similar increase in load factor, so that unit costs from nuclear generation would fall. Any improvements in capital costs and efficiency for coal plant, on the other hand, were commonly forecast to bring about only a very modest reduction in coal generating costs. Forecasters relied here on the learning curve. With a technology in its infancy, like nuclear power, there were bound to be all sorts of cheap and easy ways to improve its performance, and so performance would improve rapidly. A mature technology, like coal burning, on the other hand, has been improved so much in the past that any future improvements are bound to be difficult to find, troublesome and expensive. If we make these historically important assumptions in our hypothetical case, then it will be wise to begin building reactors before they are competitive with their coal burning rivals. Early nuclear plant may produce more expensive electricity than coal plant could, but the experience gained in its building and operation will enable later reactors to show lower costs than coal generation, the benefits after this point being more than enough to pay for the reactor's R & D costs and the high cost of early reactors.

Other factors which help to determine generation costs are not under such human direction. The most important of these are the future price of uranium and of coal, and future electricity demand. These forecasts need to be made over a sufficiently long time ahead to distinguish between the two plants, remembering that the fuel costs of later years make less contribution to lifetime generating costs than those of early years. The capital intensity and high capital costs of our hypothetical reactors means that their economic operation requires a long life, of say twenty-five years, and to this must be added the lead time of five to twelve years to give a forecasting horizon of twenty-five to thirty-five years. Thus the CEGB's calculations of the net effective cost of nuclear plant employs forecasts of coal and uranium prices thirty-one years ahead, Monopolies and Mergers Commission (1981). But is this at all realistic? Can uranium and coal prices and electricity demand be forecast over this period with sufficient accuracy to justify investment in either coal or nuclear plant? As we shall see time and again in the history of nuclear power, perceptions of the future of energy demand and supply have changed fundamentally in a very short time, often in less time than it takes to build a single reactor, let alone a whole programme of them. Time and again, forecasts of these factors over the required period of twenty-five to thirty-five years have come to be seen as grossly wrong within five to ten years. Investing in a programme of our hypothetical reactors is therefore very open to error; the forecasts which it demands simply cannot be made with the required degree of confidence.

Forecasts of nuclear capital costs and load factors present a similar problem. The capital intensity of nuclear plant means that the cost of its electricity is highly sensitive to capital costs and to load factor. But these cannot be known until the reactor is operational (Surrey and Thomas (1981)). It may be known before then that such and such building costs have been met, but from this it will be impossible to forecast with any degree of certainty what the final capital costs might be. Delays, complications, failures and newly discovered safety problems can all add to capital costs, and delays due to lack of parts, strikes etc. can increase capital costs through incurring extra interest on loans. Only very near the end of construction can any confidence be placed in estimates of capital cost. Similarly, load factors are only known when the plant has been working for a number of months or even a few years, while it is debugged. This means that the cost of electricity from a particular reactor can only be known with confidence once it is built and running. For coal–burning plant, on the other hand, there is a very long history of building and operating them, which enables confident predictions to be made about capital costs. In addition, of course, their lower capital intensity means that the cost of coal generated electricity is not sensitive to capital costs.

The long lead time of our nuclear plant means that it will take many years to acquire knowledge of capital costs and load factors, and so many years to appreciate generating costs. Suppose that the lead time is seven years. If a decision has to be made today about whether to order a nuclear or a coal-fired plant, then we know the capital costs of nuclear plants and their load factors only for those ordered earlier than seven years ago. In making a forecast of the capital costs and load factor of a reactor ordered today and operating in seven years time, this is the only data available. The forecast must be based on the capital costs and load factors of reactors which are operating today and so which were ordered at least seven years previously. In the early stages of an expanding nuclear programme there may be very few reactors operating, and if they are early ones they are likely to be significantly different in a number of ways from those of today's design. Thus learning about the way in which capital costs and load factors are changing, i.e. testing our original forecasts about declining capital costs and increasing load factors, is a very slow business. This makes for a great deal of noise in the little data which may be available. Suppose, for example, that our hypothetical reactor does not show declining capital costs over the first four units or so. What is the significance of this? It is enough to falsify the original forecast that they would show a steady fall, but how will they behave in the future? Will the early 'teething troubles' be sorted out and the expected learning curve followed from the fifth unit on, or will capital costs continue to increase or stay constant? Load factors may be considered similarly. If these turn out to be falling for the first few units, is this a sign that they will continue to fall, or will the problems causing this soon be solved with a steady rise in load factors from then on?

The problem of slow learning is worsened by the large units in which the reactor is designed and built. The fewer units which make up the nuclear component of electricity supply, the less experience is gained about capital costs and load factors; learning about nuclear costs becomes slower. Thus the high capital costs and capital intensity, lead time and unit size of the reactor combine to ensure that learning about generating costs is slow. Forecasts of capital costs and load factors for the programme of reactors take a very long time to be compared with experience, and when they are the significance of any deviations from what has been forecast are unclear. Slow learning means that the data on which forecasts of capital costs and load factors can be based accumulates very slowly, so that errors in these forecasts are likely to occur and to persist for many years.

I may be accused of being very old fashioned in talking of forecasts in this way, for it will be objected that the modern way of scenario construction is indeed a response to the recognition that these forecasts cannot be made with the requisite degree of certainty. In deciding on matters of nuclear investment, hard and fast forecasts on such things as capital costs and load factors are not attainable, but a range of values for them can be employed to see how sensitive the decision is to these factors. If the decision appears to be reasonably robust, as does for example the advantage of nuclear power over coal according to the scenarios of the CEGB's NEC calculations, then the investment can be recommended. There is certainly some sound sense in this, at least it is an attempt to accommodate human fallibility in the planning of technology. But it does not tell against the arguments I have tried to make above for two reasons. First of all, whatever sophistication of forecasting and scenarios might be used, it remains that learning about the real generating costs of nuclear stations is a slow business. The uncertainties inherent in investing in such technology are therefore slow to reduce, and so errors in investment decisions are likely to be made. The slower data accumulates the greater room there is for error. This is the point I wish to apply to the hypothetical nuclear programme, and it remains true whatever devices might be invented to improve the way we cope with uncertainties. Scenario construction is surely an improvement on straightforward prediction, and it may improve the way in which decisions on nuclear power are taken, but my point is that the degree of improvement which is possible in this way is limited by the inherently slow accumulation of data about nuclear generating costs.

The second point concerns the cost of error. In the next section it will be argued that the cost of whatever errors are made in planning a programme of the hypothetical reactors is likely to be large. If the programme is to be an acceptable risk, it follows that there must be good assurance that it is going to be successful. This is unrecognised in most scenario construction where, for example, a small range of values for load factors and capital costs is considered. For example, the CEGB's sensitivity analysis of the net effective cost of Heysham 2 AGR takes as its most pessimistic assumptions that there will be a two-year delay in commissioning, a 5 per cent reduction in load factor and 15 per cent higher capital costs than on central assumptions. As we shall see, these are very modest indeed (House of Commons Select Committee on Energy (1981), Jeffrey (1982) and Committee for the Study of the Economics of Nuclear Electricity (1982)). The cost of error clearly points to the widest possible range to be employed in the scenarios.

We may therefore draw the conclusion that in planning a programme of the hypothetical reactors, mistakes are very likely to occur, a point summarised in Figure 4.1.

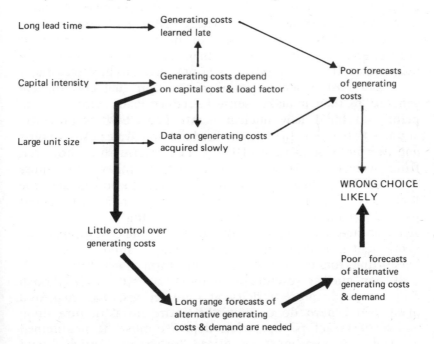

Fig. 4.1. Errors in planning the hypothetical reactor programme are likely

2. The cost of error

If mistakes are likely to happen, the next question is obviously what kind of costs they are likely to involve. The cost of erroneously investing in a reactor of the type described is four fold. First of all there is the cost of misinvestment, the electricity produced by the reactor could have been produced at less cost by coal-fired generating plant. But this is not the end of the matter. The capital costs of the nuclear plant cannot, of course, be recovered and so the logical thing to do is to continue to use it as intensively as possible to exploit its low fuel costs. For coal-fired plant of lower capital intensity, the reverse is true. If investment in a coal-fired plant intended for base load operation proves to be mistaken, it would be madness to burn coal in it as originally intended. The rational move would be to save fuel costs by reducing the load factor of the plant, in the limit mothballing it until happier days. This is what is now happening with much of the oil-fired plant ordered before 1973. This is straightforward where there is some degree of overcapacity in the generating system, allowing a shift between fuels to be made. When there is no over-capacity, however, shifting fuels means constructing new plant. If nuclear power stations can produce electricity at a unit cost which is less than the unit fuel cost of coal generation, then it makes sense to reduce load factors on coal plant and build new nuclear plant. The CEGB is presently arguing that this is the case in England and Wales (Monopolies and Mergers Commission (1981)). The reverse shift, however, from nuclear to coal is impossible because this would require the total generating costs from coal plants to be less than the fuel costs of nuclear plant. Since capital costs are the largest part of generation costs for nuclear plant, this is clearly impossible. So here is the second component of the cost of mistaken investment in nuclear reactors; the mistake cannot be retrieved by shifting from nuclear to some other form of generation.

If our nuclear reactors have to be operated even though they are producing electricity at a greater cost than coal fired plant could have done, then the entire infrastructure upon which the reactors' operation depends must be maintained as well. Uranium must be mined, processed, enriched and fabricated into fuel, the reactors' wastes need to be processed and disposed of, safety inspectors need to be trained and design teams kept in existence to cope with failures in the reactors and the associated plants, and so on. This is the third component of the cost of error. The direct costs of the operation of the infrastructure are, of course, included in the generating costs of the nuclear reactors but there are important indirect costs as well. The existence of infrastructure, even if acquired in error, can bias the next round of base generation investment by favouring reactors which can exploit the already existing capital intensive infrastructure over those which do not require it, or require further investment in infrastructure.

Finally, running the reactor on base load for its lifetime makes the next generation of nuclear plant less attractive. These too will be capital intensive, and so will require to operate with a high load factor to be competitive with coal. But with earlier nuclear plant already supplying a proportion of the base load, the scope for achieving a high enough load factor is limited. The need to live with the mistaken investments of yesterday reduces the benefits which can be obtained from tomorrow's reactors. This is the fourth component of the cost of mistaken investment in nuclear plant.

Figure 4.2 summarises the above discussion on the high cost of error in nuclear technology. Thus, not only are decisions to invest in a nuclear programme of the hypothetical reactors

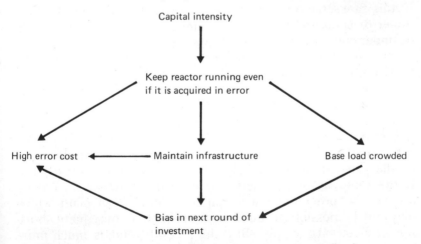

Fig. 4.2. Errors in planning the hypothetical reactor programme are expensive

likely to be in error, but if mistakes do occur they will be very expensive ones. The reactor programme has the extraordinary double curse of being open to error, and of incurring great costs if errors happen.

3. Serial and piecemeal ordering

Things look even more difficult for the hypothetical reactor if the pattern of its ordering is considered. At one extreme, reactors may be ordered *serially*, many units being under construction at the same time, so that there is a high ratio of reactors on order or being built to reactors actually operating. If more freedom to change designs in the light of construction and operating experience is wanted, orders may be placed in a *piecemeal* fashion, where construction follows experience so that the ratio of plants ordered or under construction to plants operating is kept low. Suppose 5 × 1 GW reactors, each taking seven years to build were ordered, one each year. This is clearly an example of serial ordering. If the forecasts upon which the programme are based turn out to be roughly right, then the serial programme has the advantage that the benefits of nuclear power accumulate rapidly, because all five units are operational by year eleven. But things are quite different if the forecasts turn out to have been incorrect. In this case serial ordering has the disadvantage that by the time the error is discovered, a whole number of reactors may already be under construction, or firmly ordered.

Suppose, for example, that by year six it becomes apparent that coal prices will fall steadily, quite contrary to the forecasts made in year zero, making coal generation cheaper for the foreseeable future. By year six work has begun on all five units, six years of work have been put into the first reactor, five into the next and so on until three years' work have been put into the final reactor. Two things make the abandonment of the programme in the light of the revised forecast difficult. If the forecasts of electricity demand are accurate, then abandoning the programme will call for a new fossil plant which may not be possible to build in time, with a consequent shortfall in electricity supply. But the second point is much more important. Take the first reactor, and let its capital costs be C_N, fuel costs for lifetime generation be F_N, and capital cost sunk b year six C_N^6. Economics points to abandoning the reactor (ignoring the problem of the possible shortfall in supply) if the capital costs of a new 1GW coal plant plus the fuel costs to provide the same output of electricity from coal plants is less than $(C_N - C_N^6) + F_N$. This is very unlikely to happen; by year six it is likely that so much capital will have been sunk in the first reactor that the choice will be to continue its construction. The capital intensity of the hypothetical reactors means that very soon in their construction so much capital has been sunk that it is too expensive to switch to coal generation even if it is believed that this will be cheaper. Once investment in nuclear plant has reached a certain level, it is expensive to do anything but continue its building. The capital intensity of nuclear plant therefore makes for the irreversibility of decisions to invest in them. The problem of abandoning the programme is worsened by the large size of the units, since it can only be done in 1GW steps. So serial ordering has the advantage of acquiring benefits from nuclear power quickly, if these are to be had, but involves the danger that if mistaken investments are made, the cost of error will be multiplied.

Piecemeal ordering of reactors has the advantage, unlike serial ordering, that the discovery of errors in forecasts can be used to correct the programme. In the present case, suppose it is planned to order one reactor at first, a second when the first is operational seven years on, a third when the second is working and so on. If the forecasts upon which this programme is based are found to be wrong, then the programme may be adjusted to accommodate the new information. If coal prices fall from year six to make nuclear electricity more expensive than that generated from coal, then only one reactor is being built and no more are on order. Piecemeal ordering enables the programme to be adjusted much more readily to the discovery that forecasts are wrong. The cost of mistaken forecasts is therefore much less for piecemeal ordering than for serial ordering.

The disadvantages of ordering in this way are, however, profound. First of all, if the forecasts favouring nuclear power all turn out to be substantially correct so that the caution of piecemeal ordering was mistaken, then the benefits from nuclear power are acquired very slowly. In the previous case of serial ordering, for example, there are fifty-five GW years of benefits from nuclear power by year twenty, but only nineteen GW years by this time from the piecemeal programme. Secondly, learning about the capital costs and load factors of the reactors is delayed in the same way since operating experience is built up only very slowly. Finally, the great time needed to build even a modest five reactors, thirty-five years, means that the forecasting horizon is lengthened with a corresponding increase in the uncertainty surrounding, for example, coal and uranium prices and electricity demand. It may be that before even our modest five-reactor programme is half-way completed that profound changes may occur in one or more of these factors. If this happens, then plans for later reactors can easily be abandoned but it is from these later reactors that we can expect the real benefits of the programme. Earlier reactors can be expected to be more expensive to build and operate than the later ones which can benefit from learning. Thus the early reactors may not be competitive with coal plant, serving as a learning exercise. The benefits from later plants can then be expected to pay for the uneconomic early plant and the reactor's R & D costs before this. If the programme is abandoned halfway through, however, there will be no chance to recuperate the early costs and the programme as a whole will show a loss. This point can be made in another way. Whatever factors favour nuclear plant, they may well prove to be changeable and of uncertain duration. Given this uncertainty, it makes sense to try and obtain the benefits of nuclear power as quickly as possible, before some of these factors change and the economics of generation swing back to coal. This is to favour a serial programme of reactors, and not the cautious development of piecemeal ordering.

4. Conclusions from the hypothetical case

It is now time to survey the discussion of our hypothetical reactor programme with an eye to what to look for in the real history of nuclear power in order to test Lindblom's theory of partisan mutual adjustment. The central conclusion is that nuclear power is not like ordinary novel technology. The technology of nuclear power has a long lead time, high capital costs and capital intensity, is built in large units and depends for its operation upon a complex infrastructure. These features of the technology render it highly inflexible. Capital intensity, long lead time and large unit size combine in the way indicated in Figure 4.1, to make learning about generating costs slow, which is to say that the monitor's response time is long. Mistaken investment in capital intensive plant cannot be ameliorated by switching to other fuels; correction of the error must await the wearing out of the original plant. In other words, the corrective response time is long. The need to continue with nuclear plant even if it has been acquired in error means that infrastructure must be maintained and that the base load becomes crowded, which as Figure 4.2 shows, add to the cost of mistaken investment. This is to say that the controlled error cost is high. A remedy for such a misinvestment would be to build fossil plants to replace the nuclear ones. This would take many years, during which time there would be a costly undercapacity of generating plant, and would itself be impossibly expensive due to the capital intensity of the nuclear plant. This means that the control cost is extremely high.

Inflexibility means that errors in investment decisions are very likely, much more than for decisions about other kinds of technology. Investment decisions of this sort call for forecasts of uranium and coal prices and electricity demand over twenty-five to thirty-five years, which is far too long to hope for any degree of accuracy. Forecasts of the direction in which nuclear capital costs and load factors are moving are also required, but data for forecasting accumulate extremely slowly. The same features of the technology make errors in investment very expensive. There is the cost of misinvestment itself, which cannot be ameliorated by shifting from nuclear power to other fuels. Once built, reactors have to be operated, which crowds out the base load making the next nuclear plant less attractive. Running reactors also requires maintaining the infrastructure, which adds direct costs and also provides a second source of bias in the choice of future generating plant. Serial ordering of plant gathers benefits quickly, if they are to be had, but greatly multiplies the cost of any errors in investment. Piecemeal ordering, on the other hand, reduces the costs of mistakes, but only by greatly postponing any benefits which might be obtained from nuclear plant.

If the real history of nuclear power, to be explored in the following chapters, is anything like the hypothetical case

72   *The troubled history of nuclear power*

discussed here, then it seems that partisan mutual adjustment will pass its test with flying colours.

References

Bupp, I. and J.-C. Derian (1981), *The Failed Promise of Nuclear Power— the Story of Light Water*, Basic Books, New York.
Burn, D. (1967), *The Political Economy of Nuclear Power*, Institute of Economic Affairs, London.
Burn, D. (1978), *Nuclear Power and the Energy Crisis—Politics and the Atomic Industry*, Macmillan, London.
Collingridge, D. (1980), *The Social Control of Technology*, Frances Pinter, London.
Collingridge, D. (1982), *Critical Decision Making*, Frances Pinter, London.
Collingridge, D. (1983), 'Lessons of Nuclear Power and the Future of the Breeder, I—US & UK History', *Energy Policy*, **11**, forthcoming.
Committee for the Study of the Economics of Nuclear Electricity (1982), *Nuclear Energy: The Real Costs*, Penwall, Callington, Cornwall.
Franks, C. (1983), *Parliament and Atomic Energy*, D. Phil. thesis, Oxford.
Henderson, P. (1977), 'Two British Errors, Their Probable Size and Some Possible Lessons', *Oxford Economic Papers*, **29**, 159-205.
Hinton, C. (1980), quoted in Williams (1980), p. 103, p. 245.
House of Commons Select Committee on Energy (1981), *The Government's Statement On the New Nuclear Power Programme, Vol. 1— Report and Minutes*, HC 114-i, HMSO, London.
Jeffrey, J. (1982), 'The Real Cost of Nuclear Electricity in the UK', *Energy Policy*, **10**, 76-100.
Monopolies and Mergers Commission (UK) (1981), *Report on the CEGB*, London.
Patterson, W. (1977), *The Fissile Society*, Earth Resources Research, London.
Rush, H., G. Mackerron and J. Surrey (1977), 'The AGR—A Case Study in Reactor Choice', *Energy Policy*, **5**, 95-105.
Surrey, J. and S. Thomas, 'Worldwide Nuclear Plant Performance', Appendix 5 of House of Commons Select Committee on Energy (1981), *Report on the Government's Statement on the New Nuclear Power Programme*, Vol. 3, HC 114-iii, 781-804, HMSO, London.
Sweet, C. (1982), 'Logistical and Economic Obstacles to a Fast Reactor Programme', *Energy Policy*, **10**, 15-24.
Williams, R. (1980a), *The Nuclear Power Decisions*, Croom Helm, London.
Williams, R. (1980b), 'The Structure of the UK Nuclear Industry and Its Influence on Policy', *Nuclear Energy*, **19**, 417-22.
Wonder, E. (1976), 'Decision Making and Reorganisation of the British Nuclear Power Industry', *Research Policy*, **5**, 240-68.

5   NUCLEAR POWER IN BRITAIN

This chapter attempts to analyse the history of nuclear power in Britain to see if the kind of problems indicated as likely by the discussion of the hypothetical reactor programme in the previous chapter can be found in the real world. The historical discussion relies heavily on Burn (1967, 1978) and Williams (1980).

1. Magnox

(a) History

The 1954 Atomic Energy Act set up the United Kingdom Atomic Energy Authority (UKAEA) as a public corporation overseen by a minister and funded by parliamentary vote. The new body had five functions: R & D on nuclear reactors, training of people from industry as the need arose, advising the electricity industry on matters of nuclear power, procuring adequate supplies of uranium and other materials needed for reactors and the production and reprocessing of fuel. Ministerial control would be exercised to ensure a proper distribution of effort between these tasks, but in practice control was never effective because of the absence of outside experts who could be called upon to give an independent judgement of the Authority's plans and work. The Authority was to develop prototype reactors, the next step of commercial operation being undertaken by various consortia formed by groups of companies able to provide between them entire nuclear power stations on a turnkey contract, an arrangement strongly favoured by the UKAEA. By 1956 there were five consortia. The purchaser of commercial reactors was, of course, the electricity generating industry, dominated by the CEGB covering England and Wales (earlier the Central Electricity Authority (CEA)) although reactors were also acquired by the South of Scotland Electricity Board (SSEB). The key role of the UKAEA in nuclear power development is obvious from this; not only did it have to advise the virtual monopoly buyer on nuclear matters, it also had to advise the consortia on the difficult

task of translating the Authority's prototypes into fully commercial plant. For many years it was effectively the sole
repository of technical skill and knowledge of nuclear power.

The Authority considered ideas for hundreds of types of
reactor, but underlying this was the belief that the limited
resources available to Britain called for R & D to be concentrated on a very narrow front, not dissipated over half a dozen
or more prototypes. There were at this time many constraints
on nuclear reactor design apart from limited R & D funds.
The British bomb programme having been built around plutonium, there were no large and guaranteed supplies of enriched
uranium for fuelling reactors. Natural uranium was therefore
seen as the only practicable fuel. Doubts about sustained
supplies of heavy water meant that graphite became the
favoured moderator. Unhappy about progress in America
enabling water to be used as a coolant, the British favoured
gas cooling. Plutonium for the bomb programme was made
at Windscale by two reactors fuelled by natural uranium,
with graphite as a moderator and cooled by air. Thus the
sights of the UKAEA were set and the cardinal creed of British
reactor design came to be gas cooling and graphite moderation.
Demand for more plutonium for weapons led to the building
of Calder Hall, to the so-called Magnox design, named after
the alloy used in its fuel cans. Although intended principally
to produce plutonium, the Calder Hall plant also produced
electricity for the grid, being the first reactor in the world to
do so in October 1956.

Britain's first nuclear programme was announced in a 1955
White Paper. Nuclear power was seen as a way of avoiding an
impending shortage of energy. Electricity capacity was forecast to grow from 20 GW in 1954 to 55 to 60 GW in 1975.
This would have meant increasing the coal burnt by power
stations from forty to one hundred million tons a year. The
coal industry had failed to keep up with the steady increase
in demand for coal since World War II, and this was expected
to worsen in the future. Importing large quantities of coal
or using imported oil as a power station fuel was ruled out
as placing too great a burden on the balance of payments.
The White Paper proposed a programme of 1.5 to 2.0 GW over
ten years, the early stations being of the Magnox design. It
was thought that the cost of electricity from the reactors would

be close to that from coal stations. Forecasts of capital costs were made from the Calder Hall experience, although the opening of this plant was still a year away. The nuclear reactors were reckoned to have a life of fifteen years and a load factor of 80 per cent. Fuel costs were forecast on the basis of uranium costs for the Windscale reactors, although these were reckoned to be open to greater uncertainty than forecasts of capital costs. The reactors would produce plutonium, which it was thought could be used in reactors of later design, and so a 'plutonium credit' was included in the costing in the form of a reduction in generating costs. The CEA saw the future in much the same way as the White Paper. It forecast a need to find additional fuel for 3.9 GW of electrical capacity by 1965 and 23 GW by 1981. It felt that nuclear power could not be expected to fill the gap as soon as 1965, but would be able to do so sometime after this. The CEA forecast generating costs for 200 MW coal and nuclear plants at 0.519 and 0.698d per unit respectively, the second figure falling to 0.432 when the plutonium credit was included. Thus nuclear power was seen to be soon cheaper than coal.

The 1955 programme was greatly expanded to 5.0 to 6.0 GW in March 1957. There were many reasons behind this. The energy shortage was still perceived as threatening and a larger programme would reduce coal demand even more. There was a belief that larger Magnox reactors would produce cheaper electricity, but building large plant meant a bigger programme to iron out the lumps in investment. There was also the psychological spur of the opening of Calder Hall, and the 1956 Suez crisis seemed to underline the White Paper's scepticism about relying on imported oil. The new programme was to consist of twenty stations (one in Northern Ireland). It was still thought by everyone concerned that the early reactors would be more expensive than coal generation but with steady improvements the later reactors would be considerably cheaper. Only marginal adjustments were made to the 1955 costings. The CEGB calculations done in 1958 assumed a 6 per cent rate of interest and gave generating costs from nuclear plant completed in 1962/3 as 0.63 to 0.70d per unit, compared to coal's 0.54 to 0.65d per unit.

Forecasts were, however, seen to be seriously wrong before the programme was even half finished. The impending energy shortage soon evaporated. As early as 1959 there were coal stocks of forty million tons. Only four years after the original White Paper, there was a world glut of oil and its price began to fall steadily. Electricity demand in Britain grew much more slowly than had been forecast, reaching only about 40 GW in 1975 against the forecast of 55 to 60 GW, and the 'electricity gap' disappeared.

The first few years of the programme saw its justification in terms of energy shortage disappear, but they also witnessed a dramatic change in the relative costs of nuclear and coal generation. Economies of scale began to be exploited in coal stations, so that by 1961 capital costs were about one-third lower, at 40 £/kW, than they had been in 1955. The efficiency of coal burning also increased with a shift of new stations to the coal fields. This meant that by about 1960, only five years after the original White Paper, the generating costs from coal were around 0.5d/unit, compared to the forecasts made earlier of up to 0.65d/unit. The forecasts of nuclear costs were no happier. They were pushed up by a steady rise in interest rates from 4 per cent in 1955 to 7 per cent by 1963. As more was learned about plutonium the plutonium credit gradually dwindled to zero by 1963. Capital costs of the first two reactors proved to be 25 per cent higher than the 140 to 145 £/kW which had been forecast in 1955. Load factors were heavily affected by corrosion problems, which meant that the operating temperature had to be reduced. In 1969 all but one of the CEGB's eight Magnox stations were 'derated', Oldbury by as much as 30 per cent, in recognition that this problem was to be permanent. In 1979 to 1980 the average load factor of the CEGB's Magnox reactors was 53 per cent and the following year only 47 per cent (calculated on original ratings). Two changes did, however, do something to affect the dismal picture. Uranium prices fell in the early years of the programme, but were soon to rise again, and the reactors had a longer life than the fifteen years which had been assumed in 1955.

The White Paper of 1960 recognised that there was no longer a case for nuclear power on grounds of fuel supply, this after only five years. By then, however, six stations of total capacity 2.7 GW were under construction, and two more, Oldbury and Wylfa, had been approved. The future programme was shrunk, so that only these two stations were added, to give a final capacity of 4.4 GW. The first station to send out electricity was Berkeley, two years late in 1962 and the last was Wylfa, three years late in 1971. The justification given in the 1960 White Paper for continuing with the programme was that at some time in the not too distant future, perhaps in 1970, Britain would need a third primary fuel, but for this to be a possibility, technical progress had to be made by building stations which were uneconomic in the short term.

Things continued to go badly for nuclear power in the rest of the decade. Fossil generation costs fell to 30 per cent below the 1955 forecasts by the mid-sixties, capital costs of nuclear plant were forced up still further as the CEGB's rate of return rose to 10 per cent. We have seen that the capital costs of the first two stations were 25 per cent higher than forecast, and a similar pattern emerged as the later stations were completed. The issue of capital costs was clouded for many years by the CEGB's habit of calculating historic costs, which in periods of inflation greatly understate generating costs of capitally intensive plant that has been paid for in old money. A truer picture allows for inflation, so that part of the generating costs from a plant can be put towards buying a replacement plant when this becomes necessary. When the real costs are calculated, the capital costs of Magnox plant were greatly in excess of their forecast values and the 20 per cent generating cost advantage for Magnox stated by the CEGB becomes a 30–50 per cent disadvantage. Such is the magnitude of the Magnox errors (Committee for the Study of the Economics of Nuclear Electricity (1982) and Jeffrey (1980, 1982)).

(b) Analysis

What might be learned from the early story of nuclear history? Two questions about the Magnox reactor are absolutely central: Were errors in planning the programme likely to be made? Were whatever errors might have been made likely to have been expensive?

(i) *The likelihood of error*   Deciding to invest in a Magnox programme called for forecasts about the capital costs of the reactors and the load factors they would be able to achieve, about uranium prices, about the generation costs of coal plant and, finally, about electricity demand. These may usefully be considered in two groups: capital costs and load factors in one and uranium prices and coal generation costs and electricity demand in the other. With an anticipated construction time of four years, and a lifetime expected to be fifteen to twenty years, coal generation costs, electricity demand and uranium prices needed to be forecast at least fifteen to twenty years ahead. The need to forecast this far ahead is determined by features of the Magnox reactors: their lead times and capital intensity. Reactors whose economic performance requires generation over fifteen to twenty years and which take at least four years to build obviously require a forecasting horizon of at least fifteen years. If electricity demand, uranium prices and coal generation costs needed to be forecast over fifteen years or so, we can ask whether this is at all realistic; can these things be predicted over such a period, or is the forecaster being asked to do the impossible? The forecasts made in 1955 were very wrong, but is this the fault of the forecasters or should they never have been asked to undertake such a task? In 1955 the picture right across Western Europe was one of an increasing demand for energy, especially electricity, and an impending shortage of fuel, especially coal. By 1960, it was apparent that a long period of energy surplus had begun. This happened in the time needed to build a Magnox reactor. Small wonder that by the time the Magnox reactors were beginning to deliver electricity to the grid, they had become the answer to yesterday's energy problem. The hard lesson is that forecasting what needs to be forecast to justify investment in nuclear power over the necessary time span is impossible to do with sufficient certainty. On the accuracy of demand forecasts, see Ascher (1978) and House of Commons Select Committee on Energy (1981).

The lessons here may concern forecasting, or the technology itself. Could these errors have been avoided if investment had been made to improve forecasting methods, or should people have looked with greater suspicion on technology which demanded that forecasts of this sort be made? Long-term forecasts of energy supply and demand are notoriously inaccurate because of the great speed at which perceptions of fundamental problems can change. It is just too easy to blame the forecasters for the errors of Magnox. On the contrary, we should look with great suspicion at technologies, like Magnox, which require such forecasts.

The second group of forecasts concerns the capital costs of nuclear stations and their load factors. Such forecasts cannot be finally tested until the reactor is operating because only then can the money spent on it be reckoned and its load factor be calculated during the first year of operation, although this may be only a rough guide to future performance. The load factor may change dramatically, as occurred for the Magnox reactors which were derated, but with a little luck the load factor for the first year or so can be taken as a guide to its value in the future. The Magnox reactors were planned to take four years to build, but all of them took two to three years longer. It was therefore between seven and eight years before capital costs could be known and a figure for the load factor during the first year of operation calculated. Learning about forecasts of load factor and capital costs therefore took this long. Learning was slow because of the lead time, and because of the large unit size of the Magnox stations. Nine stations were built with two reactors each, coming on line between 1962 and 1971, and so there were only nine occasions to test the forecasts, nine values for capital cost and load factor. Thus errors in the forecast so vital to the success of the reactor programme could be detected only slowly and with considerable noise. For example, the first two stations exceeded their capital costs by 25 per cent, but what did that mean for the future stations? With only two values, it was impossible to say whether future stations would exceed their forecast capital costs by the same amount, or whether they would meet forecast costs because the problems with the first two stations were just teething ones.

The planning of the Magnox reactor programme was therefore very prone to error. Forecasts of coal generating costs, uranium prices and electricity demand needed to be made fifteen to twenty years ahead, a task which with hindsight can be seen to have been impossible. Knowledge of capital costs and load factors was acquired only slowly, so that errors in the forecasts which had been made could be found only very late. The likelihood of error in the Magnox programme is thus very high and is so because of certain features of the technology itself: its capital intensity, long lead time and large unit size.

(ii) *The cost of error*   We may now turn to the second question. If the Magnox programme was open to mistakes, how expensive were these errors likely to have been? First of all, there is the misinvestment in nuclear plant which cannot be ameliorated by switching to other fuels because of the capital intensity of the reactors. This is exactly what happened in the case of Magnox plant, which continued to operate even though they produced more expensive electricity than could have been obtained from fossil stations. The size of this misinvestment was greatly worsened by the serial ordering of the reactors. By 1960 it was evident that the reactor programme was a mistake, but by then six stations were already being built and there was little that could be done to reduce the error. The programme was reduced a trifle and spread over a longer period but these were minor changes. Three of the stations were more than half finished and two more were well started. The investment issue in 1960 therefore was whether the programme should be halted with the abandonment of what had already been invested. Formally, economics point to abandonment when the total generating cost of a new coal station is less than the cost of electricity from the investment needed to complete the partially built Magnox reactors, plus their fuel costs. Because of their capital intensity, this soon ceased to be true for the Magnox reactors even in the early days of construction. Serial ordering meant therefore that by the time it was known that the reactors were a mistake, a great deal of money had already been spent on them and, because of their capital intensity, more money would have to be spent on completing them. The reason for serial ordering was the need to obtain the hoped for benefits of Magnox early, but it had the effect of multiplying the cost of error when the forecasts upon which the programme was based proved to be wrong.

The reverse is true of piecemeal ordering. If the whole Magnox programme had been planned in the most cautious possible piecemeal fashion, with no reactor ordered until its immediate predecessor was operational, what might have history looked like? It is brief and dramatic. The first two Magnox reactors took about five years apiece to build, so we may imagine that only one reactor is ordered in 1957 and comes on line in 1962. By this time it is clear that coal generation is cheaper and so no more reactors would be built. Thus piecemeal ordering would have ensured that errors were far less costly than in the case of serial ordering. But what if the reactors had fulfilled their promise and been economic? The second reactor would have been started in 1962, finished in 1967 and the final reactor would only be begun in 1997, delivering electricity in 2005. Thus, piecemeal ordering spreads out the benefits of nuclear power, if benefits there be, over a ridiculous span of time. It therefore has exactly the opposite vices and virtues of serial ordering. This, it should be clear, is, as before, a function of the technology itself.

But there is more to be said about the cost of error in the history of Magnox. The reactors had to be kept running, as they were so capital intensive, but then this meant that all the infrastructure needed for their operation had to be maintained as well. Running the Magnox reactor calls for infrastructure in the form of mining uranium ore, producing uranium metal, fuel fabrication, fuel reprocessing, waste handling and disposal and inspection and operating by a small army of experts of various types. If the reactors once built are to be run, then this infrastructure must be kept intact and operating in a reasonable healthy way. There is little option but to do this, and at a time of high hopes of continued nuclear expansion beyond Magnox this may be thought to present little trouble, but at other times it does create problems as we shall see.

The need to continue operating Magnox reactors even where their total generating costs exceeded those of fossil plant which could have been built in their stead makes the second round of nuclear energy more difficult. The size of the second programme was constrained by the fact that there are 4.4 GW of Magnox stations, which are being used on base load to exploit their low fuel costs. In the early 1960s, when the second round was being discussed, summer night load was only twelve to fifteen per cent of the winter peak, so that much of the base load would be taken up by the Magnox reactors then coming on line. The next reactors, whatever their design might be, would therefore have lower load factors than the Magnox and consequently, with their high capital costs, their electricity would be that much more expensive. With a given load factor required for competition with coal, there was an upper limit on the size of the second programme. This could have been lowered by reducing capital costs, but the extent to which this was possible was unclear. Thus the costs of errors in planning the Magnox programme was likely to be high. Mistaken investment could not be limited by switching to other fuels or by scrapping partially built nuclear plant. The reactors had to be finished and operated at base load, adding the burden of maintaining the infrastructure needed to do this and also making the second round of nuclear plant less attractive. All of this is due to features of the technology: its high capital costs and capital intensity, long lead time, large unit size and need for complex infrastructure.

The technology of Magnox reactors can therefore be seen to have been cursed right from the beginning. Not only were errors likely to be made, and learning about mistakes likely to be slow, but whatever mistakes were made were likely to be very costly ones. To say that the technology was doomed to failure from the beginning would be an overstatement, but given the horrid complications of the world which had to receive the innovation, the statement captures an important truth. This is quite independent of what organisational arrangements were made for the technology. On this analysis these are quite secondary, as will become clearer in what follows.

2. The Advanced Gas Reactor (AGR)

(a) History

The second generation of British gas-cooled, graphite moderated reactors became known as the Advanced Gas Reactor (AGR). Designs for the first prototype were begun in 1957 in an atmosphere of great optimism about the line of reactors being developed by the British. The prototype came to full power in 1963. The UKAEA thought that the principal advantage of its new reactor was that it had a lower capital cost than Magnox reactors, an advantage they estimated to be about 20 per cent. The forecasts they made in 1962 assumed a 6 per cent rate of interest, and spread overheads across a 6-GW programme. The capital cost of a 2 X 500 MW AGR station was then given as about £80/kW. Generating costs were forecast at 0.5*d*, falling to 0.46*d* if a load factor of 85 per cent could be achieved, and to 0.43 if, in addition, the reactors could be operated for thirty years. But these happy enough figures were for earlier AGRs; even better was expected of the later ones, generating costs falling to below 0.4*d*. By 1964, the Authority was speaking of generating costs of even less than 0.35*d*, far below what they imagined could be achieved by coal fired stations. The CEGB was naturally less optimistic, and in 1963 estimated AGR generating costs in the range 0.47 to 5.0*d*, but with capital costs of only £65/kW. By this time the CEGB had learned of the extra costs of the Magnox programme, over which it had only a very small say, and was determined not to let history repeat itself. The CEGB therefore delayed its ordering of AGRs, arguing that it was better to be right than to be quick. This caused problems for the five consortia formed to build nuclear stations on a turnkey basis. They all complained of severe over-capacity, and by the end of 1960 two mergers occurred, reducing the number of consortia to three. In 1963 the CEGB seemed just ready to be ordering its first AGR when American manufacturers of LWRs entered the story with lower estimates for their designs.

The ensuing stalemate was resolved by a White Paper the following year laying down that the CEGB should request tenders for its first AGR station, making it open to any rival designs which might be suitable. The White Paper hinted at a 5-GW programme, all of the winning design. The CEGB therefore requested the three consortia to tender for Dungeness B—a 1.2GW AGR station to go alongside the Magnox station there—and also asked for tenders for an equivalent LWR at the same site. By this time, the AGR prototype had been operating for only fifteen months. In all, seven tenders were received. The best AGR design was judged to be the one from Atomic Power Construction (APC), although the consortium had not had the resources or manpower to submit a fully detailed estimate. This design was considered against an American BWR and the contest published as the *Dungeness-B Appraisal* (CEGB (1965)). Capital costs for the AGR and BWR were given there as £78.40/kW and £70.86/kW (notice the decimal places). Generating costs were put at 0.457*d* for APC's AGR and about 7 per cent more for the BWR (again notice the precision). There were naturally sceptics who queried the calculations, especially because BWRs ordered in the United States seemed to bear less capital costs than the one at Dungeness, but the result of the competition came as a great relief to many.

All the reactor designs considered were for reactors fuelled not by natural uranium as the Magnox plants were but by enriched uranium. This was seen as a long-term commitment for British reactors and plans were made to reopen the British enrichment plant at Capenhurst. This had been mothballed in 1963 when demand for enriched uranium for military purposes fell. The Minister of Technology approved the plans in late 1965, the cost being put at £13½ million. Even with this expenditure, British enrichment costs were expected to be 10–15 per cent greater than American costs for some years.

APC was finally given the contract to build Dungeness-B, on condition that their design and management teams were strengthened. The design they had submitted was far from complete, and many changes were made during construction. By 1968 problems with metal corrosion, graphite stability and gas flows forced a radically different design philosophy within the dimensional constraints of the first design and the hope of using as many of the already prefabricated components as possible. Work on the boilers came to an end soon after, and there were serious failures in the two steel liners of the concrete pressure vessels. In 1967 the second AGR was ordered by the CEGB from The Nuclear Power Group (TNPG), this time without tendering. The CEGB then ordered additional stations in 1968 and 1970, giving them four stations, the final three being 1.32 GW. In addition the South of Scotland Electricity Board (SSEB) ordered a station at Hunterston B. By the time Dungeness B was supposed to be in operation, four other large AGR stations were on order. All three of the consortia had work, but their designs were very different from one another.

By the end of 1970 all the AGRs under construction were having serious problems from corrosion, insulation and turbulent gas flow and it was clear that they would all exceed the costs which had been forecast for them. Changes made to Hartlepool and Heysham on grounds of safety added considerably to their costs. By 1976, in constant prices, Dungeness B was 109 per cent over cost, Hartlepool 72 per cent, Hinkley Point B 33 per cent and Heysham 23 per cent. February of that year saw the production of electricity from Hinkley Point B and Hunterston B. Dungeness B came on line on 5 April 1983, and at the time of writing the two remaining stations are expected to be delivering power before the end of the same year. The lateness of the AGR programme was mitigated by a much slower growth in electricity demand than had been forecast. If the AGR reactors had all come on stream as they were expected to do, then inefficient coal plant would have had to be closed down because of overcapacity in both the CEGB and the SSEB. As time went on the overcapacity on both systems grew so that the delay in finishing the AGRs did not produce a shortage of electricity.

Various estimates of the cost of the mistaken investment in Britain's five AGR stations have been made. For the four in England, CEGB (1981) puts the cost at £2.5 billion, which includes IDC but makes no allowance for the cost of generating the electricity which the AGRs would have supplied if they had been on time. This, not surprisingly, is a major element in the cost of the error. Henderson (1977) includes this element and compares the cost of the AGR programme with one of LWRs, which could have taken its place. Assuming that the AGR programme would be completed, as the CEGB had promised at the time of his writing, Henderson put the total cost of the AGRs at about £3.2 billion (1975), with a loss of £2.1 billion. Burn's (1978) calculations of the programme's costs are close to Henderson's at £3.8 billion (1975). Burn (1980) updates his own and Henderson's calculations, allowing for the further delays which had occurred between their being made and 1980. The programme's cost he puts at between £8.7 and £11.1 billion (1980), between 92 and 160 per cent over the original cost estimates. As in the case of Magnox reactors, the CEGB's historic cost accounting gives a generating cost advantage to AGRs over coal plant, 11 per cent for Hinkley Point B when compared with Drax A coal station. Jeffrey (1982) has attempted to calculate the current costs, which gives Hinkley Point B a 44 per cent greater generating cost than Drax A.

(b) Analysis

Many lessons have been drawn from the trauma of the AGR, but they all concern the organisations surrounding the development of the reactor, or make more general points about administrative style and philosophy. No lessons about the technology itself have been suggested which might be extended to other investment decisions, except at the trivial level that the AGR was more complex than had been thought and more difficult to build, demanding precision engineering on a gigantic scale, and that perhaps the scaling up from the Windscale prototype was too ambitious. These are all very well as far as they go, but they are limited in that, while explaining what went wrong, they say nothing about why these errors constitute the largest misinvestment in British history. This too is the limitation on those commentators who have looked for administrative or organisational lessons from the AGR story. Some complained that the CEGB was pressured by the Government to choose the home-grown technology over imported rivals; others that the whole arrangement of sharing out orders to the consortia was wrong because they ended up working with three quite differently designed AGRs. In other accounts, the CEGB is painted as too gullible and ready to believe the good news of the *Dungeness B Appraisal*, and too quick to act on it when delay would have been more expedient. The UKAEA does not escape, of course, and has been accused of pressuring the CEGB to adopt its favoured reactor and of being far too optimistic and overconfident in its own thinking. At a somewhat deeper level, Williams (1980) sees the whole episode as a reflection of the general lack of accountability in British administration. Henderson (1977) suggests that at the root of the errors are some deep-seated problems of the British political system: what he calls decorum, unbalanced incentives, secrecy and anonymity. Decorum is seen in the British love of careful definition of roles, administrative tidiness and impersonality, all of which imply more attention to the observance of administrative procedures than to getting the sums right (unbalanced incentives). Burn (1967, 1978) also sees the British political system as incompetent to deal with complex decisions such as the choice of reactor. Some of these ideas will be discussed later when the problems of nuclear innovation in Britain and in the United States may be compared. For now, it is enough to observe that whatever the merits of these observations, they throw only a partial light on the AGR story. They might explain why various errors were made, but as to why these mistakes become so staggeringly expensive they have nothing to say. This is a serious shortcoming as the really striking feature of the AGR story, like the Magnox one, is not that errors were made—for what innovation is without mistakes?—but that the errors which were made proved to be extremely expensive.

We can arrive at an answer to this question easily enough because the features of the technology which made the errors in the AGR programme so expensive are just the same as we found to make the Magnox mistakes costly. The AGRs were more complex than the earlier reactors and much larger in the hope of achieving economies of scale, so that they took longer to build. The early 500 MW Magnox stations were scheduled to take four years to complete and the 1320 MW AGR stations five years. In practice, of course, both took appreciably longer, particularly the AGRs. The AGRs were also capital intensive, like the Magnox reactors. AGRs use enriched oxide fuel, instead of the natural uranium metal of Magnox, which makes their operation dependent on enrichment plant and special reprocessing plant. Thus the AGRs make more demands on nuclear infrastructure. We may now see to what extent these features of the AGR can explain the great costs which it accumulated.

(i) *The likelihood of error* As before it will be useful to break the discussion in two, first asking if the planning of the AGR programme was an exercise particularly open to error, and then asking what the costs of these errors were likely to have been. Taking the question of the openness of the programme to planning errors, our discussion can be brief because it parallels that of the Magnox story and of the hypothetical reactor programme considered right at the beginning. The decision to invest in an AGR depends upon forecasts of its capital costs, load factor and fuel costs compared with the generating costs of rival reactors or of fossil stations, not forgetting the need to forecast electricity demand. With a lead time supposed to be five years and a lifetime of about twenty-five years, a decision to invest in a single AGR, let alone a programme, calls for forecast to be made of electricity demand, fuel cost and the cost of alternative means of generation over at least twenty-five to thirty years. This is a very long time in planning energy provision, as the Magnox example shows. The accuracy of these forecasts over such a length of time can be expected to be very low, making errors in investment decision from this source very likely. This is what is found in the AGR story. The forecasts made in the fateful *Dungeness-B Appraisal* of generating costs from BWRs were grossly in error. As we shall see in the following section, costs quoted by United States' manufacturers at this time were greatly exceeded when plant was once delivered. Electricity demand was also grossly overestimated, and an accurate forecast would have signalled a halt to the entire AGR programme. The CEGB's AGRs were supposed to come on stream in 1971, 1973, 1974 and 1976. Between the first two dates the forecast of maximum system demand stood at around 55 GW, but the actual value was about 38 GW. The forecasts for the latter two years were over the out-turn by 27 and 29 per cent. These, it must be remembered, are forecasts of demand for only six years ahead. This is remarkable because it makes the decision to invest in the whole programme of AGRs sensitive to the accuracy of demand forecast only six years ahead. If this is so, what hope is there for any degree of accuracy in the longer-term forecast of coal costs and uranium prices?

The AGR programme also involved forecasting the reactors' capital costs and the load factors they would achieve. Here, with five years taken as lead time, learning about the real values of these factors is very slow. In practice, of course, it has been much slower because of the delays in finishing the reactors. There is no sign that the hoped for improvement in capital costs have occurred. The CEGB (1967) puts these at (1975) £270, £310 and £270 million for Hinkley Point B, Hartlepool and Heysham, all of which are of 1250 MW. Slow learning is exacerbated by the very large size of the reactors. In all, only five were ordered and built in this period, providing only five occasions to learn about capital costs and load factors. The problem which this slow learning causes will be clearer when we turn to the cost of error.

(ii) *The cost of error*   The cost of error is first of all that of the misinvestment in AGRs when other stations could have been built to supply electricity more cheaply, or where none need have been built in view of the over-forecasting of electricity demand. But the capital intensity of the reactors means that there is no way of ameliorating this cost by switching to other fuels even if this means building more power stations.

The capital intensity also explains the problems of controlling the building programme. If money is spent on partly building an AGR, this is sunk and has no effect on the consequent investment decision of abandoning the reactor and building a different station versus carrying on to complete the AGR. Soon so much money has been sunk that a switch to another type of power station could not be profitable. Once it is under construction the capital intensity of the technology makes it costly to abandon. This is worsened by the very large unit size of the AGRs. With only five under construction the programme could only be reduced in 1250-GW steps. APC may have been incompetent to build the first AGR, but to leave history at this is to forget that they were working on a very capital intensive and very large piece of equipment, which meant that their incompetence was very expensive.

The very long lead time of the reactor presents the same dilemma we have observed before about serial or piecemeal ordering. Serial ordering, as actually occurred with the AGR, obtains the benefits, if they are to be had, but multiplies the cost of any errors. In this case five reactors were on order before the first was even due to be operating, so that serial ordering multiplied the cost of error by a factor of five or so. Piecemeal ordering might have led to the scrapping of the rest of the programme once Dungeness B was built, thus limiting the costs of error. But if the forecasts on which the AGR decision were made had been accurate then piecemeal ordering would have postponed the technology's benefits for many years. Even granting a lead time of five years, the fifth reactor would have come on stream in 1991 and with historical lead times as late as 2033.

The capital intensity of the AGR means that once built it must continue to operate for as long as possible at the highest load factor it can achieve. This adds to the cost of misinvestment in the two ways which by now should be familiar. Firstly, the infrastructure necessary to the reactors' functioning needs to be maintained, which is in itself a direct cost and a distortion in the planning of future generating plant. The AGR demands more infrastructure than the earlier reactors. It requires enrichment of uranium, fabrication of oxide as opposed to metal fuel and reprocessing of waste fuel. Thus the Capenhurst enrichment plant was re-opened in 1966 at a cost of £13½ million, to provide fuel for the forthcoming AGRs. The Windscale Thermal Oxide Reprocessing Plant (THORP), which was the subject of the now famous public enquiry chaired by Justice Parker (1978), illustrates well the distortion which the infrastructure can impose on nuclear policy. THORP was designed to reprocess oxide fuel from the AGRs, but such economies of scale were supposed to exist that it was thought that the 500–tonnes per annum plant needed for the British reactors could be expanded to one handling 1,200 tonnes per year, at only an extra 20 per cent on capital costs. If such a plant is finally built and works satisfactorily, then there will be a very large surplus capacity. It is hoped to sell this to other countries, but its availability may well bias future choices in reactor design to the use of enriched oxide fuel. THORP is called for by the need to operate with the AGRs acquired in error, and to add insult to injury, the very existence of the plant biases future investment to reactors using enriched oxide fuel and against designs like CANDU. But there is a subtler bias too, because THORP makes the breeder reactor a more attractive investment. Operating the breeder requires that its fuel be reprocessed efficiently and quickly but this presents many difficulties. An essential step in learning how to do this has always been seen to be the reprocessing of thermal oxide fuel, such as that from AGRs. The AGRs therefore ease the way to the breeder reactor. Without the AGRs there would be no need to invest in THORP except as a learning step towards operating the breeder reactor's fuel cycle. Without the AGR, the cost of THORP is on the balance sheet of the breeder; with AGR it appears on that reactor's account. This is the case despite the AGRs being acquired in error. A final point is that continuing to operate the AGRs on base load must make the next generation of nuclear plant less attractive. When they are all operating, the CEGB will have 5.2 GW of AGR and 4.4 GW of Magnox working on base load. There will therefore be less room for the next generation of nuclear plant, which will have to work on high load factors because of its capital intensity. Thus the mistakes of the past make future innovation more difficult.

We can therefore explain the great problem posed by the AGR programme in the same way as for the Magnox errors. The AGRs have high capital costs and capital intensity, had a long lead time, were built in very large units and made considerable demands on supporting infrastructure. This meant that the AGR programme was prone to mistaken forecasts and that whatever mistakes did occur were likely to prove to be very expensive ones.

References

Ascher, W. (1978), *Forecasting—An Appraisal for Policy Makers and Planners*, Johns Hopkins University Press, Baltimore.

Burn, D. (1967), *The Political Economy of Nuclear Power*, Institute of Economic Affairs, London.

Burn, D. (1978), *Nuclear Power and the Energy Crisis—Politics and the Atomic Industry*, Macmillan, London.

Burn, D. (1980), 'The Cost of the AGR', *Journal of Nuclear Engineers*, Sept./Oct., reprinted in Memorandum to House of Commons Select Committee on Energy (1981), 1252-8.

Central Electricity Generating Board (CEGB) (1965), *An Appraisal of the Technical and Economic Aspects of Dungeness-B Nuclear Power Station*, CEGB, London.

Central Electricity Generating Board (CEGB) (1967), 'Submission to House of Commons Select Committee on Science and Technology', *UK Nuclear Power Reactor Programme, Report, Evidence and Appendices*, HC 381-XVII, p. 392, HMSO, London.

Central Electricity Generating Board (CEGB) (1981), Memorandum to House of Commons Select Committee on Energy (1981).

Committee for the Study of the Economics of Nuclear Electricity (1982), *Nuclear Energy: The Real Costs*, Penwall, Callington, Cornwall.

Henderson, P. (1977), 'Two British Errors—Their Probable Size and Some Possible Lessons', *Oxford Economic Papers*, **29**, 159-205.

House of Commons Select Committee on Energy (1981), *The Government's Statement on the New Nuclear Power Programme*, HC-114, HMSO, London.

Jeffrey, J. (1980), 'The Real Cost of Nuclear Power in the UK', *Energy Policy*, **8**, 344-6.

Jeffrey, J. (1982), 'The Real Cost of Nuclear Electricity in the UK', *Energy Policy*, **10**, 76-100.

Parker, Lord Justice (1978), *The Windscale Inquiry*, HMSO, London.

White Paper (UK) (1955), *A Programme for Nuclear Power*, HMSO, London, Cmnd. 4389.

White Paper (UK) (1957), *Capital Investment in the Coal, Gas and Electricity Industries*, HMSO, London, Cmnd. 132.

White Paper (UK) (1960), *The Nuclear Power Programme*, HMSO, London, Cmnd. 1083.

White Paper (UK) (1964), *The Second Nuclear Programme*, HMSO, London, Cmnd. 2335.
Williams, R. (1980), *The Nuclear Power Decisions*, Croom Helm, London.

6   NUCLEAR POWER IN THE UNITED STATES

This chapter continues the examination of nuclear history by looking at the technology's progress and setbacks in the United States. I hope to show that the same features found to plague the technology in Britain caused the same kind of problems in America, despite the very different bureaucratic arrangements made for the innovation in the two countries. The history in this chapter relies heavily upon Bupp and Derian (1975, 1981); Burn (1967, 1978); Dawson (1976); Gandara (1977); Komanoff (1981a); Mooz (1978, 1979); Perry (1977); Rolph (1977, 1979) and Collingridge (1983).

1. History

The bureaucratic arrangements concerning the development of nuclear power were very different in the United States. The United States Atomic Energy Commission (USAEC) existed to oversee and regulate the development of nuclear power under the scrutiny of the powerful Joint Committee on Atomic Energy (JCAE). The Commission's remit was to develop nuclear power under the normal competitive conditions operating in American business. The Commission therefore relied heavily on private industry to develop civilian reactor technology. In the late 1950s there were two major suppliers of nuclear plant, Westinghouse and General Electric. The former had a considerable advantage from its work with Admiral Rickover on developing a small pressurized water reactor (PWR) for driving submarines, the first prototype of which was operating by 1953. Shippingport, the first commercially sized PWR, though only 60 MW, was based on this submarine design. General Electric's version of the light water reactor (LWR) was the boiling water reactor (BWR), which was proving harder to develop than Westinghouse's PWR. General Electric therefore came to be seriously worried that what it saw as the very lucrative reactor market could be dominated by its rival. These fears were heightened after

Westinghouse's success in securing orders for San Onofre and Connecticut Yankee between 1962 and 1963.

The answer lay in boldness. In 1963 the Jersey Central Light and Power Company ordered a 515 MW BWR from General Electric for Oyster Creek. For the first time the grounds for their order were flatly commercial: General Electric had promised a cost which would make the reactor's electricity cheaper than coal generation. For the first time a utility quite unsupported by government was ordering a nuclear plant on simple grounds of its economy. So sure of their product were the suppliers, that the contract was a fixed price, turnkey one. Jersey Central believed that within five years of its start up the reactor could be stretched to 640 MW, with a capital cost calculated on this figure of a little over $100/kW. For a coal plant to compete with this coal would have to be available at 20 cents/mbtu, as against a country-wide range of 20–35 cents/mbtu, and 29 cents/mbtu at Jersey Central's own coal-fired plants. So smoothly was the reactor expected to operate that its promised load factor was 88 per cent over the first half of its thirty-year life.

These promises seemed a little wild, especially since there was no nuclear power plant of anything like this size yet working. But the optimism of General Electric drew very little criticism. Such counter arguments as did surface accepted General Electric's ability to meet its claims, but pointed out that with the stimulus of this new power source, coal generating costs might well fall to a lower level than could be reached as yet by nuclear plant.

This was soon followed by another order for a General Electric reactor for Nine Mile Point, and the reactor firm even published a price list of nuclear plant of various sizes and refinements. Alarmed, Westinghouse was quick to match its rival. Not to be left behind in what it was sure would be the long awaited lift off for nuclear power, Westinghouse published its own price list for its PWR. The advertised cost of nuclear electricity fell from between 5.8 and 7.6 mills/kWh to 4.3 mills/kWh. That this was accomplished without a second's worth of construction and operating experience still caused no scepticism from the USAEC or JCAE, who were, after all, supposed to be promoting this new technology. Westinghouse soon won its first contract on similar terms to its rival, a turnkey contract with fixed prices adjusted only for inflation. Eight other fixed-price contracts followed and the sales record was taken as proof that nuclear power had come of age. The first non-fixed-price contract was signed in 1965 and 1966 to 1967 saw what has been called 'the Great Bandwagon Market'. United States' utilities placed orders for forty-nine nuclear plants of just short of 40 GW in total. Two other manufacturers, Babcock and Wilcox and Combustion Engineering, joined the market with their own versions of LWR technology. There was naturally very intense competition in terms of equipment, price, guarantees on fuel delivery and so on. This buyers' market naturally led to a continuing downward movement of cost estimates and by the end of 1967 seventy-five stations had been ordered of a total of 45 GW. Eighty per cent of these orders had been obtained in the bandwagon period.

To the USAEC it seemed as if LWRs could now be left to private enterprise as they had become fully commercial pieces of equipment. The only function the Commission saw as continuing was the licensing of LWRs for safety. Its research and development interests turned to what were seen as the second generation of reactor, the high temperature gas reactor (HTGR) and the liquid metal cooled fast breeder reactor (LMFBR).

The suppliers of the new technology all believed that nuclear plant would produce electricity more cheaply than coal fired plant. There might be losses on some of their early orders, but this was only to be expected and they would soon be made up from later reactors, which could benefit from learning. The assumption was that the earlier demonstration and prototype reactors had yielded enough information to make confident predictions of the cost of reactors of commercial size, between 400 and 1000 MW. A key element in this assumption was that learning was bound to occur and nuclear costs were bound to fall. The chief ways of achieving this happy result were: improvements in fuel, so that completed reactors could be made more powerful; the production of standard reactors; increasing the size of reactors to achieve economies of scale and straightforward learning about the problems of building and operating the new equipment. With all these sources of learning, it was accepted by all the suppliers and many utilities and the USAEC that costs would fall rapidly. There was talk of a fall in capital costs of $25/kW per year over the first three or four years.

All of these assumptions proved to be incorrect. Westinghouse built six turnkey, fixed contract plants but made a profit only on the first, San Onofre. Things were worse for General Electric whose seven turnkey plants all lost money. By 1966 it was becoming clear that the optimism of a few years before was unjustified. Only a pair of 'standard' PWRs and two pairs of 'standard' BWRs were built as turnkey plants, so learning quickly by duplication and easing production flows was not realised. The expected 30–40 per cent increase in power ratings after two to three years was not achieved, nor anything like it. Designs could not be easily standardised and changes in design and construction schedules greatly added to costs. Instead of falling by \$25/kW/yr for three years or so, capital costs did the reverse and rose by \$75–100/kW/yr.

With hindsight none of these disappointments is surprising. Neither firm had built a plant as much as half the size of their first turnkey ones, and later orders were four times bigger than either manufacturer had taken on before 1962. The total losses to both firms on their early turnkey contracts have been estimated to be \$(1976) 1 billion, about \$75 million per turnkey contract. But things were worse than this. Under the pressure of competition the reactor suppliers began to make promises to their customers about future prices of uranium. But their forecasts here were no better than their prediction of capital costs. The price of uranium began to rise alarmingly and Westinghouse had to renege on many of its contracts. The company is being sued for \$7 billion for this breach of contract.

The turnkey contracts themselves might well mark a notable misinvestment by the reactor suppliers, but there were also heavy losses by utilities who ordered reactors on normal contracts. The utilities believed that nuclear power had at last come of age, to the extent that reactor suppliers were offering fixed-price contracts. Their misperception was strengthened by the euphoric pronouncements of the USAEC, who provided no sceptical scrutiny of the suppliers' promises. Utilities began to order nuclear plant under normal contracts. Competition between the now four suppliers forced down the prices which they quoted and made it impossible for any of them to express anything but total optimism about their product's ability to undercut coal generation. Between 1963 and 1967 fifty-nine nuclear plants in all were ordered, accounting for nearly a third of total steam electric orders for that period. As we have seen, thirteen of these were on fixed prices, turnkey contracts, and forty-six on normal terms. But the reactors ordered here could not benefit from operating experience—for there had been virtually none, and what had been acquired was of doubtful relevance. As late as the beginning of 1970, none of the reactors in this Great Bandwagon period was yet operating. What were working were the smaller plants ordered before 1965, about 4.2 GW, compared with 7.2 GW on order. But the size of reactors had increased markedly in the search for economies of scale, although there was no experience to support the existence of such economies.

After the shock of their unexpected losses on their fixed-price contracts the reactor manufacturers were still able to convince their customers that they were now on top of these teething problems, that they understood what had gone wrong and knew how to correct it. By 1966 orders were being taken for plant six times larger than the largest in operation and cumulative operating experience became an ever smaller fraction of the capacity of nuclear plant on the manufacturers' order books, moving from 50 per cent in the early 1960s to 3.5 per cent in 1967. But neither the hoped for learning nor the economies of scale were to become a reality. Capital costs of nuclear plants coming into service in 1975 were about half as much again as those for the first turnkey plants, which began producing electricity in the early 1970s, even allowing for the general increase in construction costs which exceeded the rate of inflation. The suppliers' claims that costs would soon stabilise were not met, and their cost estimates showed no sign of improving in accuracy. On average, the capital costs of all LWRs ordered in the mid- to late-1960s were about double the original estimates which had been made for them. At this time capital costs were rising by something like $40–60/kW/yr.

2. Analysis

Looking at this history from the privileged position of today the lessons to be drawn may appear to be fairly obvious ones. The rate at which reactors were ordered was clearly excessive and far outstripped the accumulation of operating experience. This happened because the pressure of competition made suppliers of the new technology make more and more optimistic promises and made them agree to fixed prices to make their confidence public. This overoptimism was not countered by the USAEC who were, on the contrary, all too ready to see the technology they were formed to serve finally become a commercial reality. The USAEC actually worsened matters by the confidence it expressed in the suppliers' ability to meet their promises. The lessons are as clear as they are unexciting; purchasers of new and largely untried technology should be much more sceptical about the promises made by its suppliers because these may be more of an answer to competition than an expression of what can really be done.

But there are much more important lessons to be drawn than this. To see why the mistakes were made is one thing, but to explain why these mistakes proved to be so very expensive is quite a different question. In the present case, the kind of mistakes which occurred seem to have been fairly unspectacular examples of the sort of errors made every day in a busy world. What makes the LWR story so interesting is not the existence of errors like these, for what story about innovation is free from such incidents, it is the extraordinary costs which these ordinary errors proved to have which is really notable. To explain these may well lead us to much more profound lessons about what kind of technology should be developed in the future. The explanation will be by now quite familiar. Decisions about LWRs were open to error and whatever errors did happen were likely to be very expensive, and this is because of features of the technology itself: its lead time, large unit size, high capital cost and capital intensity and its demands upon infrastructure.

(a) The likelihood of error

An investment decision in a LWR or a series of LWRs requires forecasts to be made of uranium prices, LWR capital costs and load factors and the cost of alternative methods of generating electricity, not forgetting electricity demand. Forecasts of uranium prices, the cost of alternatives and electricity demand must be made many years ahead because of the lead time of LWRs and their long lifetime, needed to spread out their high capital costs over a great output. As we have seen in the earlier discussions of Magnox and AGR, these forecasts over such long periods cannot be made with sufficient confidence to justify the investment decision. In addition, knowledge of the reactors' capital costs and load factors is acquired only with operating experience, and so the reactor's long lead time and large unit size mean that learning about these vital indicators of performance is very slow. If mistakes are made in forecasting load factors and capital costs, then these mistakes will not be unearthed for many years, so that many investment decisions may have been based upon the forecasts before their inaccuracy is appreciated. For these reasons, decisions about investing in LWR technology were very likely to have been wrong. History shows us that they were, in fact, wrong, but the technology itself made this unhappy outcome highly probable. We shall return to this point a little later.

(b) The cost of error

As before, mistaken investment decisions could not be salvaged by shifting to other fuels because of the capital intensity of the reactors. Once built, the LWRs had to operate at the highest achievable load factor. This increases the cost of error because the nuclear infrastructure needed for LWR operation must be maintained, and because the operation of LWRs makes the next generation of nuclear reactors, be they improved versions of the LWR or ones of an altogether different design, less attractive because the base load is already partly taken up by the original LWRs. These error costs were multiplied by the serial ordering which took place. Unlike the British reactors, there was not a single programme of LWRs, as purchases were made by a number of utilities, but many LWRs were ordered before there was a significant operating experience, and so we may speak here too of serial ordering.

Can the utilities be blamed for this? They cannot because they were caught in the familiar dilemma of serial versus piecemeal ordering. If they had been more cautious in their ordering, and had waited for the early reactors to have been in operation before ordering more, history would have been very different. The first few turnkey plants would have shown nuclear power to have been a much less attractive investment than the suppliers and the USAEC had claimed. Whether further investment could have been squeezed from the utilities it is hard to say, but they would certainly have been much more cautious. In this way the costs of the mistaken forecasts made by the suppliers would have been limited. But to say that the utilities should have done this is to use hindsight. If the forecasts had proved roughly correct, then such cautious piecemeal ordering by the utilities would have postponed the benefits to be obtained from the new technology by many years. In the hope of these rapid benefits the utilities were much less cautious, and ordered in a serial fashion. As before, this meant that by the time it was clear that the original forecasts were too optimistic, a heavy investment in LWRs had occurred, which multiplies the cost of the error.

(c) Slow learning about capital costs and load factors

The LWR provides by far the largest fraction of orders for nuclear stations, and so we may look at the hundreds of LWRs, as compared with the handful of Magnox or AGRs, to obtain a deeper insight into the problem of learning about the capital costs and load factors of reactors. Oyster Creek was ordered in 1963 but was operating only from late 1969, so that hard information about capital costs by which the promises of the suppliers might be tested only began to accumulate from then. The significance of the first items of data was very unclear. Did the unexpectedly high capital costs of the first few LWRs mean that the technology had been pushed to commercial operation prematurely, or did it merely indicate the existence of teething troubles which would soon be eliminated in later units which would show declining capital costs? With the large units then being built and the time taken for their construction, information about capital costs grew only slowly, and the clouds of uncertainty surrounding its interpretation were therefore as slowly dispersed. But the utilities were anxious to obtain the benefits of nuclear power and began to order reactors while the issue of capital costs was still unclear. This was not recklessness on their part because the long lead times of LWRs meant that ordering in a piecemeal fashion, so that orders matched and grew with operating experience, would have delayed the benefits from nuclear power, had they existed, by many decades. By 1970 when Oyster Creek was on stream, there were already many reactors on order. It is the same story as before, only heightened by the scale of the country's generating system and by competition between reactor manufacturers. By the time the errors in forecasts of capital costs had become clear and apparent to all but the staunchest defenders of nuclear power, many reactors had been built or were under construction or on order; the scale of misinvestment was very great.

As we have seen in chapter 1, as late as 1974 the USAEC was forecasting that the steady rise in capital costs of LWRs would soon stop and capital costs would stay fairly constant from then on (USAEC 1974), and Burn (1978) repeats this message even later. With the quantity of data then available, this may well have been a defensible interpretation of the facts. But by this time the investment in LWRs in the belief that their capital costs had stopped increasing was very considerable. When it became known that these beliefs were wrong, and that capital costs had continued their upward path, it was too late to retrieve any but a tiny fraction of this misinvestment.

The first truly thorough analysis of LWR capital costs was Mooz (1978). By this time only reactors whose construction began before 1968 were operational and able to be accurately costed. Those ordered since 1968 were still being built at the time of the analysis. This left Mooz with only thirty-two reactors with reliably known capital costs. The most striking result is that capital costs have risen over the five-year span of the data base at about $ (1976) 140/kW per year. He also found that there were economies of scale in constructing nuclear stations, but that they were considerably less than had been widely forecast by utilities, suppliers and the USAEC. They amount to about $0.22/kW per MW increase in capacity, so that moving from 850 to 1050 MW plant reduces unit capital costs by 7 per cent. But the small size of the data base, due to LWR long lead times and large unit size, meant that the significance of Mooz' results is unclear. Things may very well have gone very differently since 1968, remembering that at this stage in the development of the technology there was great scope for change. The problem, of course, is that any improvements will not be identified for at least ten years. What should a utility about to expand its generating capacity in 1978 have done in this situation? The past data tell of increases in capital costs over the first five years of commercial

LWRs, and of less marked economies of scale than originally expected, but this is for reactors built up to ten years earlier. The data means very little. If it could be inferred that capital costs will continue to rise and that scale economies will continue to be modest, then the utility should obviously not order an LWR. If, on the other hand, the hard work of nuclear engineers can be expected to have brought about a reduction in capital costs since 1968 and greater scale economies since 1968, then a large LWR would be the best investment. The trouble is that there is no way of deciding with any degree of certainty which way the data point; many things may have happened in the ten years since 1968 to improve LWR economics, and many other things to worsen them. The utility's choice is therefore very open to error; the LWRs high capital costs, capital intensity, long lead time and large unit size mean that there were not enough data to justify a decision to invest in an LWR in 1978. The problem can be put in an even more striking way. An LWR ordered in 1978 could benefit from knowledge about capital costs of similar reactors ordered before 1968. But a reactor ordered in 1978 will not be delivering electricity until at least 1988, making twenty years of wasted experience. The latest capital cost data available for an LWR which begins operating in 1988 is from twenty years earlier in 1968.

Since 1978, the analysis of LWR capital costs has been furthered by a second study of Mooz (1979) and Komanoff (1981a). Mooz' 1979 study was able to use hard data on capital costs from thirty-four LWRs and estimates from a further twenty. His earlier findings were confirmed. Mooz found that average capital costs, allowing for inflation in plant construction costs generally, had increased between 1966 and 1978 at $ (1978) 140/kW per year. This time, however, no economies of scale could be detected. Komanoff examined all United States nuclear reactors over 100MW capacity that achieved commercial operation between 1972 and 1978 (inclusive), forty-six units in all. Again, allowance was made for inflation of public utility construction costs. Adjustments were then made to control for effects of unit size, multiple units, geographical location, architect–engineer's experience and use of cooling towers. It was found that nuclear capital costs increased by 24 per cent per year. There was slight economy of scale, nuclear capital costs per KW being proportional to the -0.2 power of unit size, giving a 13 per cent reduction for each doubling in size, far less than predicted by the nuclear industry.

Komanoff (1981b) explains the increase in costs by the growth in regulation of nuclear plant, pointing to the high degree of correlation between increasing costs in the 1970s and the size of the nuclear sector. There is about a 50 per cent increase in capital cost with each doubling of sector size. As growing operating experience has discovered more and more ways in which accidents can occur, there has been an ever increasing need to invest in protection against such accidents. Komanoff points to the growing number of generic (i.e. applying to all reactors of a particular kind) safety-problem communications to licencees from the Nuclear Regulatory Commission in the form of a bulletin or circular. Since 1979 the rate per reactor at which these documents have been issued has been running at nearly three times the rate per reactor in the early years of the decade. This shows how growing operating experience has increased the rate of detection of safety defects in reactors. The growth in the size and complexity of the reactors, and perhaps increased scrutiny of safety issues as the result of public pressure, has led to the discovery of many new safety failings, so that the rate of detected defects per plant has risen. In addition, the growing problems of regulating reactor safety have led to the standardisation of safety controls, if only to keep the regulating bureaucracy functioning. There is also the need to improve the safety of individual plants as nuclear capacity grows, in an attempt to keep the overall risk to the public constant. This ties in with the present discussion on slow learning. If Komanoff is right and operating experience has revealed a growing number of generic safety problems, what of the early LWRs, which could not benefit from this learning? These reactors may well be judged to be unsafe in the light of operating experience acquired since their construction, but their capital intensity makes closing them down extremely expensive. Thus the costs of unsafe early plant, which is expensive to take out of service, must be added to the familiar costs of mistaken investments in LWRs. These we have seen to be the additional cost of electricity which could have been generated more cheaply and which cannot be ameliorated by switching to other fuels, the need to keep the nuclear infrastructure supporting LWRs in a healthy state and the crowding out of base load, which makes future investments in capital intensive nuclear plant less attractive.

In conclusion, it can be said that LWRs share the same problems discussed earlier and for the same reasons. Investments in LWRs are likely to be based on erroneous forecasts, and the costs of such mistakes are likely to be very high. At the bottom of this fateful combination is the technology's high capital cost and capital intensity, unit size, lead time and demands on infrastructure.

(d) Burn's analysis

The lessons other commentators have drawn have been briefly mentioned earlier, and concern not the technology, but the organisations which figured in its development. As noted earlier, these lessons may be very interesting in themselves, but they explain only why mistakes occurred, saying nothing about the great expense of errors. It is useful, however, to consider here the analysis which Burn offers in what is the only detailed comparative study of nuclear history in the United States and Britain. Burn's first book was published in 1967, when the first turnkey LWRs were being built in America. He was, like many others at the time, quite confident in the suppliers' promises that the LWR had become fully commercial. This happy picture is compared with the sorry failure of the Magnox programme in Britain, by then already clear. This difference, Burn argues, is due to the narrow front R & D strategy adopted in Britain compared to the dozens of prototypes investigated in the United States. This, in turn, is a reflection of very different attitudes towards commercial competition in the two countries. In Britain fair competition between different designs was impossible because nuclear development up to the stage of prototype was the monopoly of the UKAEA, which was further insulated from the demands of the British utilities, notably, of course, the CEGB, by the intervention of government in decisions on nuclear investment. With the UKAEA as the only properly qualified source of advice on nuclear questions, the government was bound to be biased towards its opinions in any conflict with the utilities. In the United States, on the contrary, government stood well back from meddling in nuclear energy development questions, leaving this to private

enterprise encouraged by the USAEC. This ensured fair competition between prototypes, and the successful commercialisation of the design which proved itself to be best, the LWR.

History was, of course, very unkind to Burn's analysis. As
we have seen, the early commercialisation of LWR technology
proved as much a mistake as the Magnox programme, and the
suppliers were quite unable to meet their promises on capital
costs, load factors or uranium prices. Burn was quite right
to stress the great differences in the organisation of nuclear
energy in the two countries, but both countries managed to
make very similar mistakes in their attempts to commercialise
their own choice of nuclear technology.

But Burn's conclusions have somehow managed to resist the
unhappy history of nuclear power in the United States. In his
later work of 1978, he manages to reach the same conclusions
as before. The early troubles of LWRs are skated over. The
reactor may have had a much rougher early development than
anyone thought it would, but it is at last fully commercialised
and competitive with coal, at least in some parts of America
and Western Europe. The spread of the LWR may have been
checked by the economic recession and unreasonable public
opposition from environmental groups, but this is only a
temporary hiccough in the final story. Capital and fuel costs
might be several times higher than ever imagined in the early
1960s, but to counter this fossil fuel prices had increased greatly
since 1973, and capital costs of fossil stations have also risen.
Once the economy begins to expand and the public becomes
assured about the safety of nuclear energy, LWRs will be able
to compete with fossil generation across most of the United
States and much of the rest of the world. These happy prospects
are contrasted with the nuclear scene in Britain, where the
AGR disaster had by now followed upon the Magnox one.
Export prospects for gas-graphite reactors had collapsed and
the British nuclear industry was close to ruin. This, as before,
Burn sees as due to the lack of competition between prototypes because of the early decision by the UKAEA to concentrate what they saw as their limited resources in the narrow
front of gas-graphite reactors. As before, the root of this error
is in the bureaucratic arrangements which were made for
nuclear power. Burn's lessons are summarised in three conclusions all of which confirm the results of his first analysis:

(1) to give a monopoly of research and development in an advanced
    technology to a strongly centralised authority owned, financed,
    appointed and supervised by the Government is not a means of
    promoting rapid growth;
(2) if the object is rapid technical advance . . . it is dangerous to dispense
    with variety, the possibility of comparison, the stimulus of com-
    petition between autonomous groups, and the measure of profit,
    crude and blunt though these may be;
(3) a Minister in charge of a monopoly cannot make useful judgements
    or exercise control to protect the public when he has no basis of
    comparison, and the public is particularly unprotected when the
    Minister is nominally responsible for the monopoly's programme.
    [Burn (1978), 285–6]

At this point, however, history deals its second blow. Since
1978 LWR capital costs continued to rise and load factors to
be disappointing, making nuclear power less attractive than
generation even by the increasingly expensive fossil fuels, and
concerns about safety have grown, particularly since the failure
of the PWR at Three Mile Island in 1979. The prospects for the
sort of revival in LWR fortunes envisaged by Burn had diminished steadily since the date of his writing. What is striking about
nuclear developments in the two countries, quite contrary
to Burn, is that the innovation has proved uniquely difficult
and expensive in both, despite the great differences in bureaucratic and industrial arrangements between Britain and America.
This can now be understood. Nuclear power considered as
a piece of technology has certain characteristics which ensure
that mistakes are likely to occur in the planning of it, and
that these mistakes will prove expensive if they occur. What
is true of British gas-graphite technology is equally true of
American LWRs. This explains the troubled history of nuclear
power in both countries, despite the very different approaches
which were adopted towards the innovation.

We may now stand back a little from the details of nuclear
history and ask what lessons have been learned so far. It should
be clear that nuclear technology has certain features which
ensure that whatever mistakes are made, they are discovered
late and are very expensive. These features are: long lead time,
large unit size, high capital costs and capital intensity and
need for infrastructure. Mistakes were bound to happen in an

innovation of such heroic dimensions, but these features of the technology combine to make whatever errors that do happen enormously expensive. Three of these features are actively sought by the nuclear industry. Capital intensity is customarily seen as an advantage for nuclear technology because it confers insensitivity to uranium prices; large unit size is supposed to achieve economies of scale (though see the discussion of Mooz and Komanoff above) and the search for fuel economies has led to the need for greater dependence on infrastructure such as enrichment. Whatever the reality of their advantages, it must be recognised that these three features have severe disadvantages as well. Capital intensity means that errors in investment cannot be alleviated by saving fuel. Capital intensive plant should have a long life, so that it continues to congest the base load for many years, making the next generation of plant less attractive for this period. Large units mean that learning about capital costs, load factors and safety is slowed. Finally, dependence upon infrastructure means that continuing to operate plants requires maintaining a healthy infrastructure, and that difficulties with such things as reprocessing and waste disposal can restrict the future development of reactors themselves.

References

Bupp, I. and J.-C. Derian (1975), 'The Economics of Nuclear Power', *Technology Review,* 77, 71–85.
Bupp, I. and J.-C. Derian (1981), *The Failed Promise of Nuclear Power— The Story of Light Water*, Basic Books, New York.
Burn, D. (1967), *The Political Economy of Nuclear Power*, Institute of Economic Affairs, London.
Burn, D. (1978), *Nuclear Energy and the Energy Crisis—Politics and the Atomic Industry*, Macmillan, London.
Collingridge, D. (1983), 'Lessons of Nuclear Power and the Future of the Breeder, I—UK and US History', *Energy Policy,* 11, forthcoming.
Dawson, F. (1976), *Nuclear Power—The Development and Management of a Technology*, University of Washington Press, Seattle.
Gandara, A. (1977), *Utility Decision Making and the Nuclear Option*, Rand Corporation, R–2148–NSF, Santa Monica.
Komanoff, C. (1981a), *Power Plant Cost Escalation: Nuclear and Coal Capital Costs*, Komanoff Energy Associates, KEA–12, New York.
Komanoff, C. (1981b), 'Sources of Nuclear Regulatory Requirements', *Nuclear Safety,* 22, 435–48.

108   *The troubled history of nuclear power*

Mooz, W. (1978), *Cost Analysis of LWR Power Plants*, Rand Corporation, R-2304-DOE, Santa Monica.

Mooz, W. (1979), *A Second Cost Analysis of LWR Power Plants*, Rand Corporation, R-2504-RC, Santa Monica.

Perry, R. (1977), *Development and Commercialization of the LWR 1946-1976*, Rand Corporation, R-2180-NSF, Santa Monica.

Rolph, E. (1977), *Regulation of Nuclear Power; The Case of LWR*, Rand Corporation, R-2104-NSF, Santa Monica.

Rolph, E. (1979), *Nuclear Power and the Public Safety: A Study in Regulation*, Lexington Books, Lexington, Mass.

United States Atomic Energy Commission (USAEC) (1974), *Liquid Metal Fast Breeder Reactor Program: Environmental Impact*, WASH 1535, USAEC, Washington DC.

7   NUCLEAR POWER IN FRANCE

France will soon be the only major country where nuclear power provides a significant part of total primary energy, some 25–30 per cent by 1990. At a superficial level, France seems to have mastered nuclear technology in a way which has proved impossible in Britain and the United States. This clearly runs counter to the thesis being developed here, that the great inflexibility of nuclear technology makes it a very difficult and troublesome innovation, for here is a case where the development of nuclear power seems to run as smooth a course as any major technical advance. It will be argued, however, that this view cannot be sustained. The smooth development of nuclear power in France has been made possible by quite extraordinary political arrangements which would be unacceptable in other democratic countries, and it has involved the bearing of very great and continuing risks from the technology. The history section of this chapter depends heavily on de Carmoy (1982); Finon (1982); Lucas (1979); Schapira (1982), Sweet (1981) and Collingridge (1984).

1. History

France's commitment to nuclear power is understandable; France is the second world importer of coal, the third of natural gas and the fourth of oil. The country's response to the 1973 oil crisis was a programme of energy conservation, and the acceleration of the nuclear programme, which was given the task of providing 30 per cent of primary energy by 1990. The key to meeting these nuclear ambitions was the highly centralised nature of that part of the French economy. Electricité de France (EDF), the state owned monopoly utility, seems to have made the running in this, having previously wrested control of nuclear technology from the Commissariat à l'Énergie Atomique (CEA) by forcing the abandonment of CEA's gas-graphite reactors in favour of imported PWR technology. After 1970 EDF was allowed and actively encouraged to expand its share of the energy market, particularly in space heating, so that nuclear power could more easily substitute for imported hydrocarbons. The slogan was 'tout électrique— tout nucléaire'. EDF's enthusiasm for the programme resulted from its view that the only alternative was a declining electricity market, with of course a growing dependence on imported energy. Three 900–MW PWRs were ordered by EDF between 1971 and 1972. In 1973, but before the Middle East crisis, the Commission Consultative pour la Production d'Electricité d'Origine Nucléaire (PEON) recommended an acceleration of the nuclear programme to orders for 13 GW between 1973 and 1977. This was in the light of the uncertainties surrounding supplies of fossil fuels, the growth in energy demand and the favourable economies to be achieved by scale and serial production of French reactors. After the 1973 crisis the Commission was reconvened. It now recommended that EDF be allowed to order thirteen 900–MW units between 1974 and 1975, the limit of industrial capacity, and to maintain similar levels of ordering in future years, giving a maximum of 20 GW of operating nuclear plant by 1980. In the following battles with the ministries of industry and finance, the plan was reduced a little, but EDF had clearly won the day and was to preside over a massive nuclear programme.

Framatome was the only French based company with PWR experience, acquired from Westinghouse, and so it received the early orders for this type of plant. After 1973 Framatome, assured of its monopoly and of a steady order book, expanded its production capacity to eight boilers per year from 1976. Framatome was thereby given a double advantage. Its overheads were spread across a large number of reactors and the construction of standard plant brings about 'production line' economies. Framatome and its many suppliers are only able to maintain such capacity with the assurance of future orders from EDF, a need fully recognised by EDF itself.

Between 1981 and 1982 French reactors produced 99.5 TWh of electricity (188 mtoe), about 12 per cent of the country's primary energy. There were thirty units in operation, of 21.6 GW, twenty-one being PWRs (19 GW in all). An additional twenty-five PWRs (as well as the breeder Superphenix) are under construction, totalling 27.4 GW. The future of the programme became an election issue in 1981, with some of the Socialists opposing it. After the election, government debate centred around two proposals, those of the Quilès report, which called for the reduction of PWR orders in 1982 and 1983 from the original nine to four, and those of the Hugon report, recommending six plants. The latter report received the support of the government, and with this modest reduction in the programme, the earlier fears of its severe curtailment seem to have receded. The original target of 30 per cent of primary energy from nuclear power by 1990 has been modified only slightly to 27 per cent.

Official publications give an enormous cost advantage to nuclear power as can be seen from Table 7.1. Defenders of the

Table 7.1. Comparative Costs* for Power Plants in France per Year

	Nuclear	Coal	Oil
Investment	8.5	6.5	5.6
Running costs	3.0	3.1	2.8
Fuel	4.2	15.0–20.5	43.4
Desulphurization	–	2.8	3.5
TOTAL	15.7	27.4–32.9	55.3

*In 1981 centimes.
Source: de Carmoy (1982).

programme point to the comparatively low cost of French electricity. It is also claimed that PWR capital costs in France are less than in the United States, but as Sweet points out, the comparison is difficult because of the very different time profiles. Lead times are certainly shorter (six to eight years), but capital costs have risen in France as they have in America. Allowing for inflation of construction costs, Finon found that air cooling systems had increased capital costs by 7 per cent between 1974 and 1976, and added safety features had increased them by 18 per cent. Standardisation of design is claimed to account for a reduction in capital costs of 10–15 per cent (Carle (1982)).

2. Analysis

(a) Centralisation

It is clear that whatever success the French may achieve from their nuclear commitment will be due to a centralisation of decision making probably unobtainable in any other Western country. If declining electricity sales were to be avoided, nuclear power had to be employed, but nuclear power would only be economic if the capital costs of reactors could be reduced (because of their capital intensity), and this means series production of standard reactors (something not achieved in any other country), which must be large to achieve economies of scale. Because of their size, production of reactors must be concentrated on one manufacturer. Because the benefits of series production and standardisation can only be achieved by guarantees of orders over many years, electricity demand must be manipulated so that it can accommodate an ever growing nuclear capacity. This is the logic of the monopoly utility EDF, which was pressed so successfully through the French political machine.

Indeed, the abandonment in 1968 of the CEA's gas-graphite reactors is due more to the force of EDF's nuclear logic than to economics. EDF needed to have a standard reactor and a commitment to it sufficiently strong to persuade some manufacturer to make the large investment needed for its series production. This would obviously be much easier if decisions on reactor type were EDF's, and not open to threat by rival opinions from CEA. With the abandonment of gas-graphite reactors, CEA effectively left the policy scene, and EDF was to choose reactor (PWR) and its manufacturer (Framatome). The abandonment of the gas-graphite reactors of the CEA in favour of American PWRs was not based on commercial operating experience because at the time no large PWRs were operating in America. American PWR suppliers at this time were making all sorts of rosy promises about their products, as we have seen in the previous chapter but these were unsupported by operating experience. But delay would have been costly. If the gas-graphite and LWR designs were to be developed until enough operating experience had been accumulated to identify the best, this would have delayed the nuclear programme by many years because of the by now familiar features of nuclear technology (capital intensity, lead time and scale), which make learning about its costs very slow. Realising this EDF pushed for a choice in the absence of experience, naturally arguing for the reactor which reduced CEA's role in decision making in favour of its own. This choice may well have proved a happy one, but it must be remembered just how ill-informed and arbitrary it was. Success should not blind us to the risks which were involved. As before, these risks were multiplied by the series ordering of plant, recognised as necessary if the benefits of nuclear power, and especially economies of production, were to be achieved.

Thus the French response to the problems of innovating nuclear technology which arise from its high capital costs and capital intensity, lead time and large unit size was to centralise decision making in a way which overcame these problems but which also involved very great risks. If things turn out well for the French adventure, this may well be more due to the luck of the audacious, than cautious and careful decision making.

(b) Reservations

There are, however, a number of reservations about the success of the French PWR programme.

(i) Official published costs of nuclear electricity cannot really be trusted. They do not explain the assumptions made and use historical costs which understate the costs of the capital intensive projects of the past. Little independent assessment of costs have yet been made (Finon (1982), Sweet (1981)).

(ii) A comparison of electricity prices in France and elsewhere may not be very informative. The price in France may not include the true costs of production from nuclear plant (see below); the EDF fossil plant is more efficient than, for example, the CEGB's and French hydrostations make for low electricity costs (Sweet (1982)).

(iii) EDF has expanded its sales through special 'tout électrique' tariffs to encourage electric domestic space heating. Even if this represents a true reflection of marginal costs, it does so only in the sense that EDF's nuclear investments are covered by future payments extending twenty or so years ahead. This hardly guarantees a sound financial structure for the utility. EDF is a very heavy borrower both long and short term, and as its nuclear investments accumulate there will be a growing problem of financing its debts (Lucas (1979), 60-1).

(iv) Capital costs of French PWRs have risen, and may follow American experience with consistently rising costs. If this is found to happen, there may come a time beyond which more reactors are not economic, but with serial ordering and slow learning this point is bound to be overshot, with very heavy misinvestment in more reactors. The problem will be exacerbated if load factors are lower than expected, particularly if design faults call for a derating of a large number of PWRs.

(v) The PWR programme will not show net benefits for many years, at least twenty and maybe thirty (Sweet (1978)). During this time, costs of alternative sources of electricity might fall sufficiently below those of PWRs for the whole programme to show a loss. This has occurred in the examples discussed previously and may yet happen to the French programme. This, of course, is a risk of employing such an inflexible technology

(vi) Concern has been voiced about the safety of the French PWR design, particularly about cracks in the pressure vessel. This has been dismissed by Framatome, but there is obviously a risk that operating experience will reveal a serious fault in the standard design. If this is so and if the fault is sufficiently severe, there will be the choice of abandoning the reactors already built and those under construction, with a very expensive shortfall in electricity supply, which will be very costly to overcome by constructing non-nuclear plant, or else continuing to operate the reactors and building those already begun even though they have been judged unsafe (Sweet (1981)).

(vii) In the medium term Framatome's orders are bound to be very lumpy. To achieve economies of production it has expanded its capacity to eight PWRs per year, but nothing like this rate of ordering can be maintained for many more years. There must be some time soon when substitution of fossil fuel by nuclear ceases, and further PWRs can be built only as electricity demand increases. Exports will then be needed to use Framatome's capacity and that of the suppliers. Failing this there will be the kind of problems of underordering so well known in other countries (Lucas (1979), Sweet (1981), Lönroth and Walker (1980), Ince (1982)). But the rapid expansion of nuclear plant has a second difficulty. There will be only a few PWR orders once the main programme is over, but then around 2000 the reactors of this programme will need to be replaced and therefore orders for whatever type of plant is then thought best will expand rapidly for ten years or so, only to decline as before. The cycle has the potential of repeating itself indefinitely; the lumpiness of PWR orders might generate problems with supply infrastructure for many years, a phenomenon noted in Collingridge (1980), chapter three.

The conclusion of this discussion is that, contrary to Burn, the apparent success of the French PWR programme is dependent upon very highly centralised decision making which is able to blast away the very large obstacles posed by the capital intensity, lead time and large unit size of nuclear technology. This has inescapable risks of an economic and political kind. Competition between nuclear suppliers cannot spur efficiency; competition between reactor designs cannot be allowed to slow down the nuclear programme. Come what may a market must be found for nuclear electricity; whatever errors in planning are made by EDF and Framatome are bound to be multipled by their monopoly position and without rival views of the future of the French energy system, whatever mistakes these monopolies might make will be slow to be detected and remedied. The programme may appear successful today, but it still runs the risk that economics and considerations of safety may yet turn against it, at great cost. At a political level, the cost is simply the irreversibility of the nuclear programme, even if stopping it manages to find its way to the political agenda.

References

Carle, R. (1982), 'Report of paper to the Uranium Institute Symposium, London', *Nuclear Engineering International*, October, **3**, 5.

de Carmoy, G. (1982), 'The New French Energy Policy', *Energy Policy*, **10**, 181-9.

Collingridge, D. (1980), *The Social Control of Technology*, Frances Pinter, London.

Collingridge, D. (1984), 'Lessons of Nuclear Power and the Future of the Breeder, II—French "Success" and the Breeder', *Energy Policy*, **12**, forthcoming.

Finon, D. (1982), 'Fast Breeder Reactors: The End of a Myth', *Energy Policy*, **10**, 305-21.

Ince, M. (1982), *Energy Policy*, Junction Books, London.

Lönroth, M. and W. Walker (1980), *The Viability of the Civil Nuclear Industry*, Rockefeller Foundation, New York and Royal Institute of International Affairs, London.

Lucas, N. (1979), *Energy in France—Planning, Politics and Policy*, Europa, London.

Schapira, J. (1982), 'Some Problems in the Back End of the Nuclear Fuel Cycle', paper to *Issues in the Sizewell B Inquiry Conference*, Centre for Energy Studies, Polytechnic of the South Bank, London, 26-9 October.

Sweet, C. (1981), *A Study of Nuclear Power in France*, Energy Paper 2, Centre for Energy Studies, Polytechnic of the South Bank, London.

Sweet, C. (1982), *The Cost of Nuclear Power*, Anti-Nuclear Campaign, Sheffield.

Sweet, C. (1978), 'The Costs of Nuclear Power', *Energy Policy*, 6, 41-9.

8   INCREMENTALISM AND NUCLEAR POWER AGAIN

It is time to leave the fascinating story of nuclear power to return to our main theme. The history of nuclear power has been discussed in the previous chapters only because of its relevance to theories of policy making which describe how decisions ought to be made, and it is now appropriate to consider what can be learned from this excursion into the past. A brief recap of the first three chapters will perhaps set the scene. Chapter 1 showed that synoptic rationality cannot provide a way of making decisions about the breeder reactor for two reasons: the breeder is a non-incremental change over existing technologies, and if it is developed the breeder will soon become entrenched. The novelty of the breeder means that following the prescriptions of synoptic rationality in a strict way demands the acquisition of vast quantities of information on topics of every sort. This would not only itself be expensive, but it would postpone whatever benefits might be obtained from the breeder for many decades, and perhaps for ever. There is, however, little hope of weakening the demands of synoptic rationality in the case of the breeder because of the problem of entrenchment. If it is decided, for example, to build a programme of breeder reactors before all the information demanded by synoptic rationality has been acquired and digested, then there is the danger that unexpected ill effects will be discovered once the reactors are working. If this happens when the reactors are providing a substantial fraction of electricity generation and the technology's infrastructure is fully developed, then there will be no scope for any fundamental, rapid control of whatever ill effects are discovered. The whole economy will have adjusted to the existence of the reactors, so that any attempt to impose radical controls is sure to be very slow, expensive, disruptive and bitterly contested. The only controls which will be practical are *ad hoc* ones which cause little disturbance to the technology. Thus, any compromise with the strict ideals of synoptic rationality could prove very costly. Since the breeder reactor will soon become entrenched, a decision to develop it can be justified only if there is no possibility that some ill effect has been overlooked. This demands huge quantities of information, which takes us back to the original problem because getting all this information will be very expensive and take many many years.

It was decided in chapter 3 that this is an extreme example of a generic defect in synoptic rationality. Applying the rules of synoptic rationality to any real-world decision problem beyond simple games demands large quantities of information whose acquisition may be quite unrealistic given the pressures of pocket and clock. When the ideal of synoptic rationality cannot be met, however, the theory is quite silent about what sort of compromise with reality is permissible, giving no guidance about how to choose in less than perfect conditions. This is particularly clear in the present case, where the novelty of the breeder reactor multiplies the demands on information, and where the breeder's entrenchment underlines the dangers of choosing in the absence of complete information, but the problems which this highlights are not unique to the breeder: they arise in any attempt to apply synoptic rationality beyond games and outside the confines of text books. It was therefore decided that there was no hope in trying to develop synoptic rationality in response to its inability to accommodate choices about the breeder reactor. Indeed, the only reason for devoting so many pages to the theory is to obstruct any enthusiast who might wish to conclude from the similar difficulty which incremental views of policy making have in accommodating nuclear power that synoptic rationality should be championed. This cannot be the case because synoptic rationality also runs into problems in providing a framework for choices about nuclear power. Whatever difficulties incrementalist views have in coping with decisions about nuclear power, they are to be eased by the further development of these views, and not by a switch back to synoptic rationality. But this is to run ahead a little.

Incremental views of policy making have been developed as a response to the failings of synoptic rationality, which are so well illustrated by its inability to cope with the breeder reactor. The most developed version is Lindblom's partisan mutual adjustment described in chapter 2. This account tries to provide rules for making decisions within a bureaucratic system which do not lead to excessive demands on money, time and brain power. Nevertheless, it was found that partisan mutual adjustment cannot accommodate decisions about nuclear power, for the same reasons as before, namely the technology's novelty and its proneness to entrenchment. The neat and ordered world of partisans seeking to protect their own interests and accommodating themselves to their limited pockets, brains and time by using disjointed incrementalism simply breaks down when it attempts to digest decisions about nuclear power. The technology's novelty means that it comes to the attention of so many and so diverse partisans that the usual consensus cannot be achieved. Nor can analytical attention be applied to the small differences between options, for the differences between today's technologies and nuclear power are so many and so varied. The entrenchment of the technology also limits the scope for agreement between partisans as to how to control any unexpected ill effects which might be discovered after the technology is fully developed, so that decisions cannot be seen as being serial and lose their remedial focus. Once nuclear power is providing a large fraction of electricity generation, then whatever ill effects are discovered, there is no room for any fundamental change; *ad hoc* ways of living more comfortably with the technology must be devised. Decisions to develop nuclear power soon become irreversible.

The problem here is quite different from the similar one encountered by synoptic rationality. There, the problems associated with nuclear power are particularly sharp examples of the problems which arise wherever attempts are made to apply the rules of synoptic rationality. Unlike synoptic rationality, partisan mutual adjustment sometimes gives a perfectly workable account of how policy ought to be developed, one which may not be too far from the actual policy process. It is therefore worth considering the problems caused by nuclear power in much more depth to see if our understanding of nuclear technology, or of partisan mutual adjustment, might be deepened. Chapter 3 considered two ways in which this might happen. If nuclear power is a perfectly ordinary technology, posing no risks except those which arise from investment in any other innovation, then something is wrong with partisan mutual adjustment, and it must be rejected as a normative account of policy making, at least in its present form. If, on the other hand, nuclear power could be shown, independently of partisan mutual adjustment, to be an extraordinary technological innovation posing unprecedented risks, then this would be a success for partisan mutual adjustment because it tells us to avoid the technology. It was promised to show the latter, and it is now time to see if this has been done.

In detail, it was said that the following propositions would be shown:

(i)   nuclear power is a highly inflexible technology;
(ii) investing in technology which is highly inflexible does not further an agent's own interests.

It follows that investing in nuclear power does not further an agent's own interests. In other words, nuclear power is open to such peculiarly severe risks that it is a technology to be avoided by all who wish to preserve their own interests. If this is established, then partisan mutual adjustment must be counted a considerable success. If the reverse had been shown, and nuclear power was like any other innovation, then partisan mutual adjustment would have been falsified. (i) and (ii) therefore save partisan mutual adjustment from falsification, which is a success or corroboration. It was also promised to show that

(iii) partisan mutual adjustment cannot accommodate decisions about nuclear power because of the technology's great inflexibility.

Each of the propositions will now be defended. Chapters 4–7 explore the inflexibility of nuclear technology. It is important to stress that this was done quite independently of assumptions of the correctness of partisan mutual adjustment, or of any rival view of policy making. It was shown that nuclear power has certain features which make learning about it very slow, so that mistakes in planning the technology are likely, and that they also make such errors as do occur likely to be expensive. These are, of course, its high capital cost and capital intensity, long lead time, large unit size and dependence on special infrastructure. Slow learning and expensive mistakes are signs of inflexibility. In the terminology introduced in chapter 4, nuclear power has a long monitor's response time, a long corrective response time, a high error cost and a high control cost, which together indicate a very low degree of flexibility.

Just how this inflexibility has generated problems for the technology in several countries was the topic of chapters 5–7. It has been shown therefore that nuclear power is a highly inflexible technology.

It has also been shown that any highly inflexible technology is likely to lead to the same kind of costly errors as nuclear power, so that investing in a highly inflexible technology does not further an agent's own interests, since it involves the same peculiarly severe risks. In chapter 4, for example, nuclear power was characterised in the broadest way possible: as a black box delivering electricity which has a high capital cost, is capital intensive, takes several years to construct, is built in large units and depends for its operation upon a specialised infrastructure. This broad brush description was enough to reveal its inflexibility. Other technologies might share these awkward features and so be equally inflexible; they too would suffer from the same double curse. Planning these technologies would be open to error, and whatever errors were made would be very expensive. The planned London motorway scheme, for example, would have had a very great capital cost, would have been capital intensive, taken several decades to complete, was planned in one enormous unit and was dependent on the infrastructure of motor cars (Hall (1980)). It was intended for the movement of goods and people rather than the generation of electricity, but it shares the same dangerous features as nuclear power. The arguments of chapter 4 could therefore easily be modified to show how errors were likely to be made in planning the motorway system, and that whatever errors were made would prove very costly.

A word should be said about the meaning of the second proposition, especially the phrase 'does not further an agent's own interests'. This is straightforward enough where the person concerned wants to maximise his income, for it is clear that any investment in nuclear power is unusually risky. But people want other things besides a large income. The point is that *whatever* interests a person wishes to further, investing in nuclear power is not an effective way of doing this. Nuclear power has generally been argued for on straightforward economic grounds, but it has also been suggested as a way of achieving diversity of fuel supplies. This case for nuclear power will be examined in chapter 12, where it will be shown that exactly the same features of nuclear power which make it an unattractive economic investment also make it an extremely clumsy way of achieving the goal of a diverse fuel supply. Indeed, they make it a clumsy way of achieving anything that might be wanted. The technology's inflexibility means that learning about it is very slow, and yet its novelty means that much needs to be learned before it can be known that nuclear power is, in fact, a way of achieving some stated goal. Slow learning implies that errors are likely in planning the technology, for whatever purpose this is done, and what errors are made are likely to be expensive.

It remains only to examine the connection between the inflexibility of nuclear power and the inability of partisan mutual adjustment to accommodate choices about nuclear power, in particular the third proposition which claims that the technology's inflexibility creates the difficulty with partisan mutual adjustment. In chapter 2, the discussion was couched in terms of the non-incremental nature of the innovation of nuclear power and the inevitability of its entrenchment. If these features can be shown to be a reflection of nuclear power's inflexibility, then it is this inflexibility which is at the heart of the failure of partisan mutual adjustment to accommodate decisions about the technology. There are many novel aspects of nuclear technology, perhaps more than in other important technological innovations, but the matter does not stop here. What its novelty means is that there is a lot to learn about the new technology, but this by itself is not a problem, provided learning can be rapid and inexpensive. But, of course, for nuclear power the reverse is true, its inflexibility means that learning is extremely slow and liable to great expense.

The shift from vacuum valves to transistors might be counted a non-incremental one, for it led to many profound changes in industry and not merely to a cheaper product (Braun and Macdonald (1982)). But the new technology was produced in millions in very small units, by plant which was, at least in the early days, deliberately labour intensive and there were no problems with infrastructure. Learning about the new technology was therefore extremely rapid and cheap compared with the value of the products being made. There may or may not be more things to learn in the case of nuclear power, the question is hard to quantify, but what is certain is that the inflexibility of nuclear technology meant that learning was very slow and very expensive. The problems caused by nuclear power's non-incrementality for partisan mutual adjustment are therefore reflections of the technology's great inflexibility. The same is true for whatever difficulties are created by entrenchment. The capital intensity of nuclear power, for example, means that it is very expensive to do anything but run reactors on base load whatever ill effects the technology might prove to have, a feature worsened by the existence of capital intensive infrastructure. Thus the inflexibility of nuclear technology prevents decisions about it from being accommodated within the prescriptions of partisan mutual adjustment, and this will be the case for other inflexible technologies as well.

There are three conclusions to be drawn at this stage. First of all, it must be noted that partisan mutual adjustment is corroborated in passing the test proposed for it here. If nuclear power had proved to be a quite ordinary technology, the inability of partisan mutual adjustment to accommodate it would have to have been taken as a falsification of Lindblom's theory. Since nuclear power has turned out to be a very extraordinary technology, partisan mutual adjustment has resisted falsification and so must be counted, to this extent, a successful theory. Secondly, the inflexibility of nuclear power means that it is a technology which should not have been developed, and one whose future development should be constrained as much as possible, even if this means phasing it out altogether. This is a topic which will become clearer in the following chapters.

Finally, the difficulties experienced in the development of nuclear power can be expected to be matched by any technology of similar inflexibility. If we wish the efficient policy machine prescribed by partisan mutual adjustment to operate in a particular area of choice, then we must find ways of detecting and avoiding inflexible technologies, for the machine only works when this is done. This is no real hardship because it has been shown quite independently of partisan mutual adjustment that investing in such inflexible technology will be very risky because errors will be both likely and expensive. We therefore need ways of identifying inflexible technologies sufficiently early in the policy process for their development to be avoided. How this can be done is the subject of the remainder of the book. To prevent accusations of using hindsight the breeder reactor, which as yet has no commercial history, will be considered and attempts made to assess its flexibility.

References

Braun, E. and S. Macdonald (1982), *Revolution in Miniature*, 2nd edition, Cambridge University Press, Cambridge.

Hall, P. (1980), *Great Planning Disasters*, Weidenfeld and Nicolson, London.

PART III

TOOLS FOR CONTROL

9 CONTROLLING THE BREEDER—FLEXIBILITY IN STRATEGIC CHOICE

Part III of this work aims at generalising the results so far obtained about the particular technology of nuclear power. The normative theory of decision making, partisan mutual adjustment, is unable to provide prescriptions for the making of decisions about nuclear power. Despite this, the theory is to be retained; indeed its inability to accommodate decisions about nuclear power is a success for the theory and not grounds for its rejection. Decisions about nuclear technology are simply not the same as decisions about ordinary technologies. Nuclear power is very different from the normal run of technical innovations because of its very great inflexibility, which means that planning investment in nuclear plant is an undertaking peculiarly prone to error, and that whatever mistakes are made are likely to prove extraordinarily costly. To put it another way, learning about nuclear power will take a very long time and is likely to be very expensive. Its inflexibility also prevents nuclear technology from being subject to the normal policy processes of partisan mutual adjustment. There are very few controls which can be applied to the technology. However offensive it may prove to be to some group of partisans, the process of innovation cannot be reversed and the only controls which are possible are piecemeal, superficial and *ad hoc*. Any more radical changes which some partisans might favour are sure to be too expensive or slow to operate, so that partisans must adjust to the technology, not the technology to the varying interests, values, factual discoveries and power relationships of the partisans.

This is obviously not a problem restricted to nuclear power. Many other technologies have properties which make them just as inflexible and unamenable to control through partisan adjustment. Large international airports, for example, are built in very large units, have high capital costs and capital intensity, have long lead times and a great dependence on infrastructure, such as road and rail links with nearby cities and a local town to accommodate workers. These technologies can be expected to be as troublesome as nuclear power. They can be expected to be open to planning errors, and whatever errors do happen can be expected to be expensive. At the same time their inflexibility will mean that they cannot be controlled through the normal policy processes involved in partisan adjustment. As before, partisans will have to adjust to the technology and not the technology to the partisans. In other words, there is a *general* problem about many technologies of which nuclear power is but one example. So far, however, this is the only case which has been examined. What is now needed is the development of a decision methodology which can prevent the kind of problems encountered in the development of nuclear power from occurring with other technologies. Controlling a technology requires preventing the build up of resistance to change as the technology matures and diffuses, becoming increasingly woven into the whole technical and social fabric. Preventing the development of this inflexibility means that the technology is open to control through the normal process of policy making, because the technology may be altered as the partisans involved continue their never-ending dance of mutual adjustment. Inflexibility, on the contrary, means that there is very little scope for the partisans to adjust, so that they have to make do with the status quo plus or minus *ad hoc* controls, which are the only kind available to them.

A methodology is therefore needed which can show how the flexibility of technologies may be measured, and at which stage in the policy process this can and should be done. It should also identify what responses are appropriate to the discovery that a technology has a low flexibility, and how flexibility is related to other decision methodologies, such as cost–benefit analysis and utility analysis. Part III attempts to address all of these issues through the example of the breeder reactor. This is appropriate for a number of reasons. It obviously fits in with the earlier discussion of nuclear power, of which the breeder is merely a special case, and the breeder has been discussed already in the first chapter. Breeder technology is also very topical. Breeder reactors have for many years played a central role in the energy policy of many industrialised countries. In Britain experience has been gained with the small Dounreay Fast Reactor and the 250–MW Prototype Fast Reactor, also at Dounreay, which has been producing electricity since 1975. The first commercial sized breeder reactor, the

Superphenix, is being built in France, where completion is expected in 1984. In the United States, President Reagan has attempted to revive the smaller Clinch River Project, which was mothballed by President Carter, and development work is also underway in the Soviet Union.

The importance of breeder reactors is seen as lying in the different efficiencies with which thermal and breeder reactors consume their uranium fuel. A breeder reactor can obtain fifty one hundred times the energy which a thermal reactor can obtain from a given quantity of uranium. Against this, breeder reactors are more expensive to build than thermal ones. It is, therefore, necessary to trade off this extra expense with the savings from the breeder's efficient use of fuel. The world envisaged for large breeder programmes is one where uranium reserves have been seriously depleted by thermal reactors already built or under construction, so that it is an expensive commodity. It should be noted that known uranium reserves are just about adequate to fuel all the world's present thermal reactors throughout their lives. It is possible that large high-grade reserves will be found in the extensive areas of the world so far unprospected, but there is no guarantee of this. It is also thought that by the time large breeder programmes are possible, oil reserves will have been seriously depleted so that oil cannot substitute for expensive uranium. Difficulties are also foreseen in any attempt to increase coal production sufficiently to substitute for uranium; thus the only way to ensure adequate supplies of energy is through the efficient use of uranium, which results from its consumption in breeder reactors. A second advantage of breeder reactors is that they produce electricity directly, and demand for this extremely versatile form of energy was expected to increase greatly until the early years of the next century. The conventional view is well put in a Department of Energy publication:

the trend of demand for electricity and doubts about the availability of fossil fuels (and other alternatives) on a sufficient scale point to the need for a large and increasing nuclear component in our energy supplies by the turn of the century.

. . . the United Kingdom has no important uranium resources of her own, and if it remained totally dependent on thermal reactors it would become increasingly vulnerable to the world price and availability of uranium. Fast reactors would reduce the impact of increases in uranium prices and reduce the possibility of our not being able to supply our demand for uranium from the world market. It seems essential, therefore, to keep open the option of using them. [Department of Energy (1975)]

A further reason for considering the breeder is that it is still in the early stage of development. As we shall see, it is here that considerations of flexibility are especially important. Finally, it is a technology which on a superficial view threatens to be very inflexible, and so is a candidate for closer inspection in this regard. This may explain a curious puzzle. Electricity accounts for only about 20 per cent of Britain's final energy, a figure which is fairly typical for similar economies, but discussions of energy policy very often reduce to talk about electricity supply. Talk about electricity generation seems inevitably to centre on nuclear power, and discussions of the latter more often than not become concerned solely with the breeder reactor. This is because the things which create problems in the energy system, problems which are quite disproportionate to their role in providing energy, are technologies which are inflexible. Much of the energy system shows commendable flexibility, as shown, for example, in the replacement of solid fuel by gas in domestic heating; the substitution of natural for town gas and of smokeless fuel for coal and in the dieselisation and, later, electrification of the railways. Such changes have proved to be quite straightforward and to pose no extraordinary problems, which demand the attention of energy policy analysts. In contrast, nuclear power, which is a very minor contributor to UK primary energy and far less to the world's has had to receive enormous political and analytical attention.

1. Flexibility in strategic choice

Two levels of decision making about technology are to be distinguished: strategic, which is the topic of this chapter, and tactical, which will be considered in chapter 11. Strategic choice is the selection of what *kind* of technology is to be developed in the medium-long term, say from ten to fifty years. Strategic choice therefore concerns the direction in which R & D is to be conducted. It is not the selection of a detailed and well researched technology from a set of similarly investigated technologies, but the decision to conduct the R & D which will eventually enable such choices to be made. At the stage of strategic choice, very little is known about the details of some of the technologies being considered, so that only guesses can be made about their capabilities and drawbacks, and their appeal to today's decision makers and to tomorrow's as well.

The aim of strategic choice should be stated as widely as possible. 'Meeting future demands for useful energy' is thus a better statement of aim than the narrower 'meeting future demands for delivered electricity'. A prematurely precise aim may rule out many possibilities far too early in the decision process. Armstrong (1983) argues that the United Kingdom Royal Commission on the siting and timing of the third London airport was misled in this way. Its brief was to find the best site for a new four-runway airport to serve London, and to determine when the new airport should be built. It was therefore barred from considering other ways of meeting demand for flights, such as the expansion of a single existing airport in the region, or the expansion of the national system in a piecemeal way or the shifting of transit movements to some foreign airport. One of these might have been superior even to the best sited new airport, but the Commission could not explore this possibility for purely bureaucratic reasons. This is therefore an example of strategic choice narrowed by administrative fiat. The selection of the *kind* of solution to the growing demand for flights, whether the building of a new airport, the expansion of one existing airport, the expansion of the whole system in some way or the shifting abroad of some demand, is a typical strategic choice. Once it is made, then tactical decisions, such as choosing the best site for the new airport, finding the best airport to expand, finding the best way to increase the capacity of the whole airport system or the best place to shift demand, may be made. The error of the Royal Commission was to be forced to consider tactical questions before any thought had been given to the strategic ones.

The stated aim to which a strategic decision is directed may be met by many dozens, hundreds or even thousands of technical *options*, each of which is a realistic candidate for meeting the aim, but they can be organised into a handful of *strategies*. A strategy is simply a set of options which appear capable of meeting the stated aim, which are variations on a theme. If the aim is to meet future demand for international travel, then one strategy is to build a new large airport. The options within this strategy are the various sites for the airport with the various starting dates for their construction. A second strategy is to extend one existing airport, and its constituent options consist of the candidates for expansion, with the different timetables for construction which seem appropriate. A third strategy consists of the much larger set of options for expanding the airport system as a whole, and a fourth is the set of foreign airports to which demand might be shifted. Some strategies are mutually inconsistent. If, for example, a brand new airport is being built somewhere, it makes little sense to worry about what foreign airports might take excess transit movements, and vice versa. But this is not the case for all strategies: some will be clearly compatible with each other, and the relationship between others may be difficult to clarify. Shifting transit passengers to foreign airports, for example, seems obviously compatible with the strategy of *ad hoc* expansion of the entire airport system, since some demand might be met by shifting transit movements abroad and some by expansion of the home system.

Comparison of options within a single strategy is relatively straightforward. In the first strategy for meeting future demand for international travel, for example, sites for the new airport may be compared in great detail, perhaps to the extent of performing cost–benefit analysis, such as Roskill (1971), or applying utility theory (De Neufville (1976)). This is obviously dynamic. As more and more research is done to investigate the various short-listed sites, comparison between them should become easier. Comparison of this sort is possible because the difference between sites is *incremental*. Building the airport on one of two sites will involve a great overlap in planning, which means that attention can be directed to the relatively small differences between the options. It may be very difficult, for instance, to compute absolute figures for capital costs or noise nuisance, but the difference in capital cost and noise nuisance between the sites may be much easier to find. It is similar for the surface links required, and the operating problems of the sites and so on. This may be contrasted with attempts to compare options which are taken from different strategies. This proves to be very much more difficult, even to the point of impossibility because the options are different in a non-incremental way. Consider, for example, building a new airport at a specified site with the option of shifting demand to a particular foreign airport. The cost and benefits of the first are qualitatively different from those of the second, so that no overlap exists which would make for incremental comparison. Instead, all the costs and all the benefits of the first option must be compared against all the costs and all the benefits of the second.

Enough has, I hope, been said to show the great importance of strategic choice, and also to expose the very great problems which are inherent in it. In making tactical decisions, considerable knowledge has generally been acquired about the options being evaluated, and the familiar tools of policy analysis may be applied in order to aid the selection of the best option. But this knowledge is the result of R & D, which has been done because of some strategic decision taken in the past which identified this set of options as worthy of investigation and development. This strategic decision was, therefore, by its very nature made in the absence of detailed knowledge about all of the options it encompasses. Here then is the paradox: the very reasons which make strategic decisions so important make it necessary to take them in a state of extreme uncertainty. The problem is to see how choices of this kind, which are at the same time so important and so ill-informed, should be taken. I hope to show that flexibility holds the key to this. Even at this early stage, when detailed information about the options within a strategy is not yet available, it is possible to assess the flexibility of the options, and in choosing between strategies this will be a central factor because strategies consisting of highly inflexible options should be dropped.

It is worth saying a little more about each of these claims. In chapter 4 the flexibility of a hypothetical thermal nuclear reactor was considered and found to be very low. All that needed to be known about this reactor was that it had a long lead time, was built in large units, had high capital costs and capital intensity and depended for its operation upon special infrastructure. Flexibility was assessed by response time, error cost and control cost. This was enough to reveal the great inflexibility of *any* nuclear plant of whatever design, provided that it shared these features. All options which involve the construction of some such nuclear plants, whatever the details of its type and design, will be highly inflexible options. This was shown without the very detailed knowledge of the variety of reactors and their performance and fuel costs and environmental hazards, which would enable cost–benefit calculations to be done to select the reactor with the lowest generating cost, or to choose between nuclear and fossil plants. The flexibility of options can be assessed on far far less information than is demanded by a calculation of costs, so that flexibility can be measured at a very early stage in the decision process, long before the information needed for costing has been acquired. This claim will be defended in more detail in chapter 12.

If a strategy consisting of inflexible options is called an *inflexible strategy*, then I am suggesting the decision rule that there is a prima facie case for rejecting inflexible strategies in strategic choice. This now needs to be defended. As argued before, it is generally impossible to perform a detailed comparison of options from different strategies because the exercise expands to a size which defeats human knowledge and intellect, the options being non-incrementally different. For this reason it is impossible to justify the choice of one option over its rivals; there can never be enough information for this when options from different strategies are being considered. Another way of saying this is that any choice of one option from a set of options drawn from different strategies may be in error; i.e. if more facts about the various options were known, then the policy maker would revise his original preference and favour a different option. For the way facts can alter preferences in this way see Collingridge (1982, 1983). If the risk of error cannot be eliminated, then it makes sense to choose in such a way that if mistakes are made they may be discovered quickly and easily, and that the original decision can be revised quickly and cheaply if such discoveries are made. In short, flexible options should be favoured. An inflexible strategy contains no flexible options, and so should be rejected prima facie in making a strategic choice—I say prima facie because there may be some special pleading for the strategy to show that it ought to be followed even though it is inflexible. Popular arguments of this sort, some of which will be considered in the next chapter, are that the strategy is inevitable and must be followed, sooner or later, or that it is needed as a hedge against the worst which a highly turblent world might do. The condition does not make the rule empty, for without some such special pleading a strategy found to be inflexible is to be rejected.

This may be related to the discussion of the hypothetical thermal reactor of chapter 4. It was argued that this reactor and an equivalent coal burning plant were different in a non-incremental way, thus there was no hope of justifying the selection of either one for electricity generation. Though the nuclear plant cannot be shown to be the best investment, for this would mean comparing it with all other options, there are discoveries which would show that it was not the best, i.e. would show an investment in the technology to have been mistaken. If, for example, capital costs for the plant are much higher than expected, or if load factors are lower, or if it proves much less safe than originally forecast or if the expected growth in electricity demand fails to occur, then the nuclear plant is shown not to have been the best investment. The problem, of course, was that the discovery of these errors takes a very long time and that there is little that can be done to revise the decision once it has been shown to be wrong. In short, nuclear power was seen to be highly inflexible. Errors are likely to occur in planning nuclear investment, and whatever errors do occur are likely to be very expensive. To safeguard against the sort of troubles discussed in Part II, which plagued the development of nuclear technology, inflexible technologies should be rejected, and hence inflexible strategies should be dropped in making strategic choices.

The point may be made in a slightly different way in terms of learning about technologies. Strategic choice concerns selecting those technologies which are worth learning about, even though very little may be known about them at the time when such a choice must be made. It seems clear therefore that strategic choice should favour technologies which may be learned about quickly and inexpensively. These are, of course, flexible technologies. It will take many years to learn about the costs and benefits of an inflexible technology like nuclear power, and the learning process is likely to be as costly as it is slow. Since there are no variations on the nuclear theme which are flexible, it is, at least prima facie, a technology which is not worthy of further investigation. Putting it formally, since there are no flexible options within the strategy of nuclear power, the strategy may be condemned as inflexible and should be abandoned unless a special case can be made out for it.

A final point is that the flexibility of a strategy, once measured, does not remain constant over time. If a strategic decision is made to conduct R & D into a strategy, then discoveries are bound to be made which alter the original estimate of flexibility, so that there is feedback between strategic choice and the development of various options within the strategies. For example, nuclear power plants have been found to take much longer to build than originally forecast, so their flexibility, though never high, has further diminished.

2. Strategic choice and the flexibility of the breeder

Remembering the earlier warning about premature precision in the statement of objectives, the aim to which breeder technology is directed may be taken as that of meeting future demand for useful energy. The problem to be considered here is the strategic choice of technologies which might be able to meet this aim in, say, fifteen to fifty years, and in particular whether breeder reactors should be developed for this purpose.

One strategy to be considered is obviously the building of breeder reactors, and the options within this may be taken as the various types of breeder, introduced at various dates and built at various sites. Other strategies which are candidates might include the construction of convertor reactors; the construction of coal fired generating plant; the utilisation of wind, solar or tidal energy and improving the efficiency of energy conversion, transmission and end use so that demand can be met with no major investment in new supplies, the conservation strategy. The first thing to show is that these options for meeting future energy demand do, indeed, form sets of strategies, where comparison of options within the same strategy is far easier than comparison of options taken from different strategies.

The problem of comparing thermal nuclear and coal generation has been discussed in chapters 1 and 2. The two technologies are so different that there is no overlap which would enable the comparison to be directed at the small areas of difference. Instead, the difference is non-incremental, so that the entire spectrum of costs and benefits of one option must be compared with the whole set of costs and benefits for the other, as shown in Tables 2.1 and 2.2. In the discussion in chapter 1 of the USAEC's attempted cost–benefit analysis of the breeder reactor, it was similarly observed that breeder reactors differ so markedly from thermal reactors, such as LWRs, that, again, a comparison is non-incremental, including all the costs and benefits of each plant. Table 2.1 gives the elements of cost for electricity generated by a thermal plant. Table 9.1 gives the same for the breeder reactor.

Table 9.1. Elements of Cost for Breeder Electricity

Direct

Capital costs of breeder reactors
Load factors of breeder reactors
Price of natural uranium
Breeding gain of the breeder reactors
The inventory of breeder reactors
Out of reactor time (ORT) for plutonium recycling
Reprocessing efficiency
Reprocessing costs
Breeder fuel fabrication costs
The cost of decommissioning breeder reactors
The cost of waste management
The plutonium supply from thermal reactors
The cost of plutonium from thermal reactors

Indirect

Cost of accidents at breeder plants
Cost of health damage from normal emissions from the breeder fuel
    cycle
Proliferation of nuclear weapons
Loss of civil liberties to safeguard nuclear material

The breeder is even more capital intensive than thermal reactors, so its generation costs are even more closely tied to its capital costs and load factor. The rate at which fresh plutonium is produced is determined by the breeding gain, which is a feature of the reactor itself and likely to change over time, by the out-of-reactor time of fuel being recycled and the efficiency of reprocessing. The fuel cycle costs of fabrication and re-processing also need attention, as does the quantity and cost of the plutonium available from thermal reactors. The capital intensity of the breeder means that generating costs are fairly insensitive to fuel cycle costs, but they need to be forecast as a check on the investment decision. Breeders will have higher capital costs than thermal plants, and this must be offset by their low fuel cycle costs. It if looks as if the breeder's fuel cycle costs are going to be close to those of thermal reactors, then there is no reason to incur the heavier capital costs of the breeder.

The rate at which plutonium is produced from the programme is conveniently measured by the linear doubling time, being the time taken for enough plutonium to have accumulated to fuel double the capacity of the original reactors, allowing for plutonium being stored and recycled. Figure 9.1 shows how this is a function of ORT, breeding gain and reprocessing efficiency. The doubling time needs to be forecast to give an idea of the timescale over which benefits from the breeder will accumulate. A long doubling time means continued reliance upon (presumably) more expensive thermal reactors for many years, as a means of meeting electricity demand and of producing plutonium which can be used to fuel breeders to accelerate the otherwise slow rate of substitution. These relationships have been much studied in the literature. Chow (1980), reviews fourteen analyses of the United States' breeder programme in addition to USAEC (1974) and ERDA (1975). See also Vaughan and Farmer (1976), Nicholson and Farmer (1980), Sweet (1982) and Finon (1982). The availability of plutonium from thermal reactors, and its cost also need to be known, but discussion of this is best postponed for the present.

If Tables 2.1 and 9.1 are compared, similar entries will be found under the indirect costs, but they refer to quite different items. The fuel cycles are very different, and will have different health costs from their normal emissions and from accidents. Similarly the problems of proliferation and civil liberties are likely to be much more serious with breeder technology since plutonium will be transported, processed and traded much more than previously. The same may be said of reprocessing, fuel fabrication, decommissioning and waste management costs in the direct costs, which will be quite different for

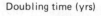

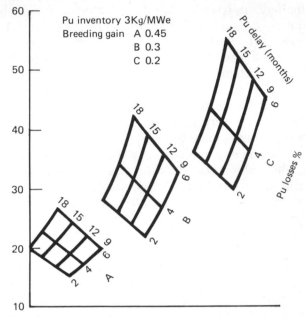

Fig. 9.1. Effect of fuel cycle parameters on breeder doubling time. (*Source*: Vaughan and Farmer (1976).)

thermal and breeder reactors. Convertor reactors, like the high temperature gas reactor (HTGR), were also briefly mentioned in chapter 1, where, again, they were found to require non-incremental comparison with thermal reactors, breeder reactors and coal generation. Recent attempts to compare the benefits of investing a given sum in thermal generating plant and in conservation measures reveal that these too are non-incrementally different (Department of Energy (1983a, 1983b)). In summary, the various options for meeting future demand for useful energy do fall into strategies where incremental comparison is possible for options within a single strategy but options from different strategies are non-incrementally different. This rules out any attempt to make detailed cost–benefit calculations comparing options from different strategies, despite the brave attempts which have been made such as the one discussed in chapter 1. Nevertheless, the flexibility of such options can be measured at this stage in the decision process.

We may therefore turn to consider the flexibility of breeder technology. It is clear that breeder reactors will be of higher capital cost than today's thermal plants. There is no reason to suppose that construction times will be reduced; on the contrary, the added complexity of breeders is likely to increase lead times. There is even more need to achieve economies of scale for breeders than for thermal reactors because of their greater capital costs, so the unit size of commercial breeders is likely to be very high. The infrastructural requirements of breeders are considerably greater than those of existing reactors. The economics of reprocessing thermal oxide fuel have always been the subject of debate, and the operation of thermal reactors does not demand reprocessing. Breeders, on the other hand, can only operate at all if their spent fuel can be reprocessed safely and speedily, enabling the bred plutonium to be extracted to fuel future reactors. Special fuel fabrication plants will also be needed and waste disposal is likely to be more difficult. Most scenarios for the introduction of breeders also assume the operation of thermal reactors for a considerable period to provide plutonium for the breeders before their capacity is enough to make possible a self-sustaining breeder programme. In this way the breeder demands the further support of thermal reactors, thermal fuel reprocessing plant and supplies of enriched uranium. To this may be added the political infrastructure designed to guard against proliferation of nuclear weapons. In short, the breeder, whatever the details of its design, will have all those features of existing reactors which make mistakes about their development both likely and expensive, only more so. It will have a longer lead time, larger unit size, greater capital costs and intensity and it will be more demanding of infrastructure. The breeder will be even less flexible than thermal reactors. Errors of investing in breeder reactors will therefore be even more likely and will be even more costly. Let us see if these fears are justified.

(a) The likelihood of error

The following discussion will be confined, for the sake of simplicity, to the direct economic costs of the breeder although indirect costs will be the subject of a later section. If the breeder is an investment of unacceptably high economic risk, as I wish to argue here, questions such as the safeguarding of civil liberties and the safe disposal of wastes may be put aside, since a second line of attack is not needed. This means that what will concern us is the list of items under direct costs in Table 9.1. If the cost of electricity generation by a breeder reactor needs to be known, then forecasts of a huge number of items are needed. As in the case of thermal reactors discussed earlier, the plant's capital cost and load factor and uranium prices all need to be forecast as does future demand for electricity. But much more than this is needed, and this illustrates the important lesson that as a technology's dependence on infrastructure increases, the likelihood of planning errors increases. The point is simply that if the performance of the technology and of its supporting infrastructure both need to be forecast, this is open to more errors than making forecasts about the technology alone.

In the present case, the operation of the breeder reactor requires reprocessing of fuel and material from the so-called blanket of the reactor where uranium 238 is transformed into plutonium. Reprocessing is needed to extract the plutonium which has been created in the reactor so that it can be used to fuel more breeder reactors. The performance of reprocessing is crucial to the economics of the breeder, and as we have seen it depends upon the breeding gain in the reactor itself, the reactor's inventory of plutonium, the time needed for fuel to be reprocessed and put back into a reactor (ORT) and the efficiency of the reprocessing plant in separating the plutonium, for some is always bound to be lost. Thus planners of breeder reactor programmes have the additional burden of having to forecast the performance, and, of course, the cost of reprocessing. This involves forecasting the breeding gain, plutonium inventory, ORT, reprocessing efficiency and the cost of reprocessing. Another part of the infrastructure will be special fuel fabrication plants because plutonium is much harder to handle than the uranium fuel for thermal reactors. The time needed for fabrication of fuel needs to be known, since it is an element in ORT, and the costs of fabrication must also be forecast. The final entries in the direct costs of Table 9.1 refer to the availability and cost of plutonium from thermal reactors, a discussion of which is best postponed for the moment.

Having seen what forecasts are needed to calculate the direct cost of generating electricity from breeder reactors, let us turn to how accurate they can be expected to be. The benefits from a breeder programme are acquired only after a number of years, and before embarking on a programme there must be some assurance that electricity demand and uranium prices will continue to favour the breeder at least for this period of time, enabling the early costs to be balanced by later benefits. This calls for a forecasting horizon of at least thirty to forty years, but we have seen that this cannot be achieved for electricity demand and uranium prices. Forecasts over this period of time just cannot be made with sufficient confidence to justify the investment. Decisions based on such forecasts are therefore very open to error. Thus the cost benefit of the breeder discussed in chapter 1 (USAEC (1974), ERDA (1975)) had a time horizon of forty-five years. The analysis of the US National Academy of Sciences (1979) looked forward thirty-five years, and Richels and Plummer (1977) sixty years. A number of similar studies, all with time horizons of this sort, are reviewed in Chow (1980).

We have seen how the lead time, capital intensity and large unit size of thermal reactors leads to a very slow learning about their capital costs and load factors, consequently errors based on too optimistic forecasts of these are very likely to happen. The same is true of the breeder, only more so. Its capital intensity makes its generating costs more sensitive to capital costs, the need to compete with thermal plant is likely to demand very large units and lead times are certain to be greater. It will therefore take many years before enough data on capital costs and load factors are available to make reliable forecasts. In the meantime forecasts of these are likely to be in error, particularly those which postulate a learning curve for capital costs, such as Keck (1981), USAEC (1974), ERDA (1975) and see Finon (1982) for official French optimism about declining capital costs. The same problem besets attempts to forecast breeding gain and the plutonium inventory needed for a breeder reactor. This requires operating the reactor, thus data accumulate just as slowly as for load factors and capital costs.

Suppose that five 1.5 GW breeder reactors are to be built, one being ordered every year, with a lead time of ten years. It is forecast that capital costs will fall throughout the series as experience accumulates, so that a learning curve is followed.

Let us consider the problem of testing this forecast. If the first reactor is built to time, its capital costs will be known by year ten. If things go equally well, later reactors will be completed in years eleven to fourteen by which time there will be data on five reactors. If these show that capital costs have increased, the original forecast is clearly wrong, but what is the significance of this for further orders? After fifteen years there are still only five data points, making it still very difficult to forecast what will happen to capital costs in the future. Mechanisms can be imagined, like that proposed by Komanoff (1981), which increase capital costs. Similar mechanisms, like the traditional learning curve, can be imagined which would do the reverse. Which of these is going to be dominant is impossible to tell with the very limited data base available. Whatever forecasts are made of future capital costs, they are therefore likely to be wrong. The effect of unit size is clear here. If the programme's 7.5 GW were made up of 0.5 GW breeder reactors, ordered three per year, then capital costs for fifteen plants would be known by year fourteen. If the lead time of these small plants is less than that of the large ones, say seven years, then data on capital costs of fifteen plants would be available by year eleven. The sooner this kind of information is available the earlier it can be used to improve forecasts of future capital costs. The same point about the capital costs of breeders can be made about their load factors, breeding gain and inventory. Learning about these will be slow for the same reasons.

Linear doubling time also depends on ORT and reprocessing efficiency. Reprocessing plants will themselves be capital intensive and will show economies of scale favouring large plants with consequent long lead times. As before, these factors will mean that learning about the breeder's fuel cycle is a very slow business. The first breeder fuel reprocessing plants are likely to be fairly inefficient, with a lengthening of doubling time. But if they are built in large units, experience will accumulate only slowly. This is especially so because their capital intensity means that once built, they should be used for as long as possible, and that abandoning an early primitive plant for a more efficient later one will be very expensive while the former still has any kind of working life left to it. Since learning will accumulate slowly, forecasts of the operating efficiency and costs of future reprocessing plants are likely to be in error, even though they may be crucial to a decision whether to invest in a breeder programme. The same point may be made about ORT, which depends on a whole number of factors about which very little is yet known. The same slow learning is bound to affect forecasts of fuel fabrication costs because this plant is also capital intensive. In short, forecasts of breeder reactor fuel cycle costs are likely to be erroneous. For recent estimates of fabrication costs see Keck (1981) and Finon (1982). Some idea of the capital intensity of breeder reprocessing plant may be obtained from looking at thermal reprocessing plants. The British THORP plant has a capital cost estimated at £800 million for a throughput of 600 tonnes per year and a working life of ten years (Monopolies and Mergers Commission (1981), 183 ff.). Estimates for two plants planned in Germany are DM 20 billion for 1,400 tonnes per year and DM 4 billion for 350 tonnes per year. Comparable figures also exist for French plants (Finon (1982)).

It can be seen that planning investment in breeder reactors is even more open to error than planning thermal reactors. To put it another way, there is more to learn about the technology of breeder reactors than about thermal reactors. The breeder's capital cost, load factor, plutonium inventory and breeding gain all need to be learned, and their dependence on reprocessing requires learning about ORT, reprocessing efficiency and costs and fabrication costs. The nature of the technology means that learning about all these factors will inevitably be very slow, data will accumulate at a very slow rate. When learning is slow, errors are likely.

(b) The cost of error

Having seen that the breeder reactor has features which make planning for its introduction open to error, we may now consider what the costs of such errors are likely to be if they are made. An error here means investment in a breeder reactor (or its associated fuel cycle plants) which produces electricity throughout its life that could have been produced more cheaply by a thermal reactor of some design or by a fossil generating plant. As noted before, this places considerations of proliferation, civil liberties, waste disposal and even plant safety to one side. If the breeder is not economically attractive, then there is no need to become embroiled in these tangled questions.

There is first of all the cost of misinvestment, which the additional capital intensity of the breeder reactor makes even less open to alteration by way of saving fuel cycle costs and passing generation to other fuels than we found in the case of thermal reactors. Thermal reactors once built are very costly if not used, and this will be even more so for the breeder because breeders will be more capital intensive and their construction will be accompanied by the building of specialised fuel fabrication and reprocessing plants, which will also be capital intensive. If an investment in breeder technology proves to have been a mistake, there will be little that can be done to alleviate the costs of the error. Even where the forecasts upon which the investment decision are based are recognised as seriously wrong during the plant's construction, there will be little room for responding to this sad discovery. As with thermal reactors, but even more so now, the capital sunk into a partially built breeder reactor would mean that it soon becomes very expensive to do anything but carry it through to completion. A programme of breeders will therefore be even more irreversible once begun than an equivalent programme of thermal reactors.

The second element in the cost of error is that the infrastructure needed for the breeder's operation must be maintained. The breeder's capital intensity calls for the greatest possible output which can be achieved, even though the reactors may have been acquired on the basis of erroneous forecasts. But operating one or more breeders calls for a whole infrastructure of expensive plant, more than is demanded for thermal plant. In particular, fuel fabrication plants and reprocessing plants are required, which will be more complex than similar plants for the thermal fuel cycle because of the large amounts of plutonium to be handled. Reprocessing is of course necessary for the operation of the breeder, unlike thermal plants where it is an option which depends upon economics. The maintenance of the infrastructure has a double cost, as has been observed previously. It has direct costs such as the capital and operating costs of the various plants associated with the fuel cycle, and this obviously counts towards the breeders' generation costs. In addition, there is the indirect cost of a bias which affects future energy investment. Fabrication and reprocessing plants will be capital intensive and are always

spoken of as having marked economies of scale. It will therefore be very tempting to build large plants to exploit these
economies, larger than required for the immediate breeder
reactor programme. Thus the French PURR plant is scheduled
to reprocess fuel not just from the Superphenix, but from this
and the next four breeders. The plants' capital intensity will
mean that little is saved by not using them, and so in the next
round of ordering for base load plant the breeder will have
been given an edge over whatever rivals it is being considered
against. Even if the first round of breeder reactors has been
built in error, subsequent rounds of investment will be biased
to favour continued investment in breeder reactors.

The third aspect of the cost of error is the congestion of
base load. Even if breeder reactors are built in error, they have
to be operated at as high a load factor as possible, even more
so than thermal reactors, which have a lower capital intensity.
But this is bound to make the next generation of capital intensive base load plants less attractive because these too need to
operate on high load factors.

It can be seen, then, that not only are breeder reactors
more open to errors in planning than thermal reactors, but
whatever mistakes do occur in planning breeder investment
are likely to be even more expensive than corresponding errors
in the planning of thermal plant. To put it another way, learning about breeder technology is not only bound to be slow,
but it is bound to be hugely expensive as well. The problems
caused by the technology's dependence on infrastructure in
both making errors likely and costly should be clear. The cost
of error, or the cost of learning, is compounded by the problem of how to order reactors, to which the discussion now
turns.

(c) Serial versus piecemeal ordering

The planning of the breeder programme is bound to be beset
by the problem of how to order its constituent reactors. At one
extreme serial ordering will ensure that the benefits of the
reactor are obtained, if the forecasts on which the investment
decision are based turn out to have been correct. But if the
forecasts are wrong, and the breeder represents a misinvestment,
then the costs of the error are multiplied. Most writers of
scenarios concerning the breeders' introduction recognise the

advantages of serial ordering, but are peculiarly blind to its dangers. Thus, in their cost–benefit analysis of the breeder reactor, USAEC (1974) forecasts that breeders would be introduced from 1990, and would have a combined capacity of 500 GW by 2000. Assuming a ten-year lead time, all of this plant must be ordered without a single second of operating experience of a commercial breeder reactor. To see what risks this entails, suppose the worst to happen, that none of the reactors work. Taking the capital cost of a contemporary thermal plant as about $1,000/kW, and assuming an advanced reactor to cost half as much again, 500 GW of unworkable plant represents a sunk investment of about $750 billion, easily the biggest misinvestment in history. Some of the investment would be recoverable by building fossil or thermal steam plant to supply the existing turbines, but these ordinarily take ten years or more to build, and would take longer because of the very lumpy ordering that would be necessary. It is much more likely, however, that the advanced reactors will work, but that their electricity will be produced at greater cost than other methods of generation. If capital costs of about $1,500/kW make the new reactor competitive, then if it is found to cost $2,000/kW, the 500 GW represents a cost of $250 billion. If this were to happen then, as before, the capital intensive plant would continue to operate, and savings on fuel costs would not be possible. The infrastructure needed for the reactors would have to be maintained, and the reactors would make the next generation of capital intensive plant less attractive.

If it is thought that I am picking on one scenario, the same point may be made about many other studies. Vaughan and Farmer (1976) talk of breeders being introduced in 1987 with a world contribution of about 250 GW by 1997, most of which would have to be ordered with no operating experience at all. Marshall (1981) sees nothing extraordinary in a scenario where breeders are introduced from 2005 with a world-wide capacity of 250 GW by 2015. Marsham (1977) seriously discusses a scenario for the UK where the ratio of reactors on order to reactors operating is 55 to 3 in 2000. Nicholson and Farmer's (1980) scenario sees the first breeder in 1990 and a saving in uranium equivalent to 80 GW of breeders by 2000. The United States National Academy of Science (1979) considers a number of scenarios for the introduction of advanced reactors. A typical one assumes introduction from 1987, with a capacity of about 110 GW ten years ahead in 1997. The same cavalier attitude to operating experience can be found in scenarios from the World Energy Conference (1978) and Häfele (1973).

Piecemeal ordering, of course, has the converse properties. If the forecasts on which the investment in breeder technology was made are wrong, piecemeal ordering limits the cost of error because only a few plants are on order at any time, but if the forecasts are correct, this cautious form of ordering delays the benefits of the breeder for many decades. To take an extreme example to make the point, piecemeal ordering under the sort of scenarios studied by the United States National Academy of Sciences (1979) might have meant only 1 GW of breeder reactor ordered in 1990. This would be operational by 2000, when further investment in the technology would be considered. If the breeder is a disappointment, perhaps it simply refuses to work, then piecemeal ordering has avoided the huge misinvestment which serial ordering would have brought about, discussed in the previous paragraph. But if the forecasts on which the breeder was ordered are accurate, so that it represents the cheapest form of generating electricity, then by 2000 there is 1 GW when there could have been 500 GW. In this case the cautiousness of piecemeal ordering has led to a severe misinvestment in non-breeder generating plant. With inflexible technology like the breeder reactor, misplaced caution may be just as expensive as mistaken optimism.

In practice, some reasonable middle path between the extremes so far discussed is obviously called for, although, as we shall see, it is not easy to find. To consider this, another scenario might be useful, the one from the UKAEA to the Royal Commission on Environmental Pollution (1976), representing the maximum British nuclear programme, subject to the constraints of uranium supply and plutonium production in thermal reactors (Table 9.2). It is worth remembering, if only to be reminded of how perceptions of nuclear energy have changed over so few years, that a footnote to the scenario states that it marks a 'considerable reduction' over earlier figures from the UKAEA. A little imagination is called for here, as it is now impossible to have a first commercial breeder by 1985, but this does not alter the point to be made. A ten-year

Table 9.2. UKAEA Breeder Reactor Scenario

Year	Breeder capacity
	GW
1980	0
1985	1
1990	4
1995	19
2000	33
2005	61*
2010	89
2015	145*
2020	200
2030	370

*Interpolated.
Source: Royal Commission on Environmental Pollution (1976), 179.

lead time for commercial-sized breeder reactors will be assumed, which is generous and less than the lead time now typical of United States' thermal reactors. The scenario shows the same alarming features as the ones mentioned earlier. In 1984, 16 GW of fast breeder reactors will be ordered in the complete absence of operating experience. By 1989, the first IGW is operating, but 29 GW are on order. After that, the ratio of capacity ordered to capacity operating drops to 3.2 in 1994, and 1.75 in 2009. If the risks involved in this pattern of serial ordering are to be reduced, then breeder capacity must be expanded following operating experience. The first problem in trying to do this is that operating experience increases in a very lumpy way, so that if orders are to follow experience, they too will be lumpy, with very inefficient use of construction capacity.

What will the lumpy introduction look like, and what problems will it generate apart from the inefficient use of construction capacity? Suppose the first IGW of breeder is ordered in 1975 and is operational in 1985. This is a risky venture with only experience of prototypes, a single plant is therefore reasonable. But how many plants can reasonably be ordered in 1985 on the basis of the first IGW? There would obviously be room for argument here, but the points I wish to make are qualitative, and not quantitative, so let us take an order of 3 GW. This is operational by 1995, but routine ordering would still be very rash with such little operating experience. Suppose, then, that 9 GW is ordered in 1995, keeping a capacity ordered to capacity operating ratio of three. If these are successfully operating by 2005 let us suppose that there is then sufficient experience for serial ordering, which proceeds at the rate of 9 GW per year, using the construction capacity of 1995 efficiently. This gives a breeder capacity of 22 GW by 2015, far less than the 145 GW envisaged by the UKAEA's scenario. By 2025, breeder capacity would be 102 GW, instead of the UKAEA's suggestion of 285 GW. Whatever the actual figures, it is clear that the need to expand breeder capacity in line with accumulated operating experience means that the ordering of the first few rounds of the programme will be very lumpy, and that programmes will take considerably longer to reach a given capacity than often recognised by the technology's supporters.

Delay generates two problems. The breeder's economics are governed by the relative cost of capital and enriched uranium. If it is worth investing the extra capital in order to save on the enriched uranium needed by thermal competitors, then the breeder reactor is to be chosen. But the market for uranium is extremely complex and open to some quite bizarre fluctuations, which cannot be forecast. For this reason any cost advantage of the breeder is likely to be shortlived and to be of very uncertain duration. Given that the early reactors are likely to be more expensive than thermal reactors, then there is a good reason to compress the reactor's development so that the programme as a whole shows a profit before uranium prices change against it. Slowing the programme down in the way we have seen to be necessary makes the whole programme much more risky; it may well happen that uranium prices will no longer favour the breeder before the programme shows a profit. The second problem caused by the delay takes us back to the previous section. It means that learning about the breeder and its fuel cycle is slowed even further. The first problem requires more detailed attention.

(d) *The uranium price signal*

The previous sections show that the breeder reactor is a highly inflexible technology, which means that whatever errors occur in planning a breeder programme are likely to be very expensive. The breeder becomes economic when prices for fossil fuels are high and when the price of uranium for use in thermal reactors is also high. A given capacity of breeder reactors will have higher capital costs than the same capacity of thermal reactors or fossil burning plants, but this is offset by their very much lower fuel costs, since they will use uranium so efficiently. In talking about the timing of the breeder, it is usual to imagine a future where thermal reactors are considerably cheaper than fossil plant, so that base load generation gradually shifts to the former. As the growing capacity of thermal reactors eats into the world's limited reserves of uranium, poorer and poorer uranium ores have to be exploited, and the price of uranium is driven up. There will eventually come a time, it is argued, when the price of uranium will be so high that it will be worth bearing the extra capital costs of breeder reactors in order to use the fuel more efficiently. After this, breeder reactors will begin to substitute for thermal reactors just as they originally displaced fossil base load generating plant.

Just when this happy day will come about depends on a number of unknowns, particularly the rate at which new high grade uranium reserves are discovered and the speed at which thermal reactors take over base load generation. If the substitution of thermal reactors is rapid, and additional reserves hard to find, then breeders will soon become economic, while a slow rate of substitution and the discovery of plentiful reserves will obviously delay that event. There is, however, no need to forecast these very uncertain factors because the price of uranium will act as a signal indicating when it is economic to start building breeders. Thus the US National Academy of Sciences ((1979), p. 124) reckon that a breeder with capital costs 40 per cent greater than those of a LWR would be economic were the price of uranium to go above $(1976) 110/kg. Nicholson and Farmer (1980) suggest that if uranium prices were to move above four times their 1979 levels, then the same breeder would be economic. The inflexibility of breeder technology means that any mistiming of investment will be very expensive. To guard against this, a sound indicator is obviously required to signal the beginning of breeder investment. The question we may now ask is whether the price of uranium is a sound indicator. If it is not, then mistiming is a danger, and this is sure to prove extremely costly.

The factors which determine the market price of uranium are extremely complex, something which is quite forgotten in the simple story about the steady rise of uranium prices until they make investment in the efficient breeder economically worthwhile. To begin with, there is no such thing as a uranium market, most purchases being under long-term contract and the spot market being used to balance short-term supply and demand. Nor do we find the usual multitude of buyers and suppliers which economists are so fond of. There are only a handful of buyers and supply is concentrated to an extreme degree; the United States, South Africa, France (including Gabon and Niger) and Canada produce about 90 per cent of the world's uranium, giving a geographic concentration unequalled by any other mineral (Radetzki (1981), Buckley *et al.* (1980)). One company controls all of South African production, and the French *Commissariat à l'Énergie atomique* (CEA) controls production in France, Niger and Gabon. In all, six decision units control up to two-thirds of world production. The picture is further complicated by the need to enrich natural uranium before it can be used in most of the world's reactors, increasing the proportion of uranium 235 from 0.7 per cent to 2 or 3 per cent. There is nothing like a market in enrichment, the USAEC still possesses a virtual monopoly. This, as we shall see, greatly complicates the pricing of uranium. In addition, the reprocessing of spent fuel obviously enables uranium to be used more efficiently, tending to reduce demand and hence the price of the metal.

From 1968 to the end of 1973 uranium prices were depressed and fairly stable, within the range $5.50 to $7/lb for long-term contracts and about the same on the spot market. In 1974 an explosive increase in prices occurred, taking prices from about $7/lb in late 1973 to above $40/lb in mid-1976. Prices stayed at this level for a number of years, although inflation meant that this represented a steady decline in real terms. Radetzki (1981) shows that this explosion is quite unprecedented.

Other commodities have had similar or even steeper increases in price, but this has always been followed by a rapid collapse, quite different from the case of uranium. The only parallel is petroleum. This price increase was completely unforeseen. In the late 1960s producers were expecting a boom in nuclear power and hence in uranium, and therefore found themselves in 1973 with considerable overcapacity and overproduction, leading to growing stockpiles of the metal. Low-cost reserves of uranium were also perceived as more than adequate to meet any likely increase in nuclear capacity throughout the world. Not surprisingly, all the forecasts in the industry at this time were for the continuation of depressed prices, with, at best, slow and modest price increases. Radetzki therefore tries to explain the extraordinary behaviour of the price of uranium. She points to a number of factors which were involved.

There was at this time a growing realisation that reprocessing of LWR fuels would be much slower in coming than originally expected. In April 1973 President Carter announced an indefinite embargo on exports of reprocessing technology and a ban on domestic reprocessing, in response to fears of the 'plutonium economy'. This obviously led to an increase in forecasts of demand for uranium, with, of course, an upward push on prices. The second factor was American policy on enrichment. In 1973 the USAEC introduced new terms for enrichment under which buyers would have to enter into 'fixed commitment contracts', promising to buy fixed amounts of natural uranium at fixed dates and receiving back at fixed dates enriched uranium from eight to eighteen years after signing the contract. The need for such long term and inflexible arrangements meant that buyers rushed in to secure future supplies, fearing that they would otherwise be squeezed out. At this time, it must be remembered, forecasts gave a buoyant future for nuclear power. The oil crisis of 1973 was a third factor, although the rate of substitution of nuclear for fossil plant and the overall expansion plans for nuclear power remained very little affected by the sudden increase in oil prices. The final major factor isolated by Radetzki was the operation of a cartel, always a danger where supply is so concentrated. It is known that the Canadian government was the central party in an international price-fixing agreement between 1972 and 1975, and seems to have caused prices to increase at the critical time by restricting supplies. In addition, there were a number of minor factors operating in the same direction: delays to Australian mining, the French withdrawal from the uranium market, Westinghouse's delivery failure and the non-proliferation policies of the United States all helped to push up uranium prices.

What are the lessons of this piece of history? The complexities of the uranium market mean that quite extraordinary and completely unexpected changes can occur in the price of uranium, of the same order as the 1973 increase in oil prices, which have very little connection with supply and demand. This means that uranium price is far from the clear beacon needed if investment in breeder reactors is to be correctly timed. The price signal, on the contrary, is bound to be very obscure and open to a number of interpretations, some of which will point to beginning breeder programmes and some which will not. If the price of uranium exceeds that thought to make the breeder economic, how can the proponents of the new reactor be sure that this is a reflection of long term supply and demand so that the price will be sustained for long enough to justify entry into the new technology?

As we have seen in the previous section, uncertainty about the permanence of a high uranium price might seem to indicate a rapid expansion of breeder capacity so that its benefits can be achieved while the technology is favoured by the price of uranium. But this would be to run into the problems discussed earlier about serial ordering of breeder reactors. If, for any reason, the decision to begin a crash programme turns out to be wrong, then serial ordering multiplies the cost of error enormously. But, this apart, a crash programme of breeders is really quite unrealistic. It takes something like between ten and fifteen years to develop new uranium reserves, and perhaps a little longer for countries with less experience in mining and surveying. If uranium prices are pushed up to a level where breeders are thought to be economic by non-market factors, such as a cartel of suppliers, it might therefore take ten to fifteen years before new supplies reach the market, forcing the price down. Within this period, a huge programme of breeders could have been started, only to find that the price of uranium has turned against the technology even before the programme is half finished.

(e) Social costs of the breeder

The novelty of breeder technology ensures that it is likely to have a variety of indirect social costs, which have not been discussed until now. The argument so far has concentrated on the direct, economic cost of the breeder because these are surrounded with less emotion than the social costs and because if it can be shown that the breeder is not an attractive investment when economic costs are taken alone, then there is no need to be concerned with indirect costs since they can only worsen the evaluation of the technology. The purpose of discussing these indirect costs now is merely to show that the problems surrounding them are symmetrical with those that have been found to surround the direct costs. As an example, consider the problem of the proliferation of nuclear weapons. Many opponents of nuclear power argue that the breeder will lead to many smaller countries acquiring nuclear weapons by diverting plutonium from their breeder reactors into the manufacture of bombs. Against this, proponents of the technology argue that there are major technical problems in making such weapons, which can easily be worsened by such devices as making stored plutonium highly radioactive with some suitable contaminant. They also believe that international agreement and inspection will be a sufficient deterrent to any government who might wish to manufacture atomic bombs.

The argument between these two parties has now gone on for many years with, perhaps predictably, no sign of consensus or, indeed, of any significant changes of mind. The conclusion from this should be that in the present state of our knowledge about nuclear technology and political behaviour, it cannot be predicted whether the breeder will promote the proliferation of nuclear weapons. The question therefore becomes how long will it take to learn whether or not the breeder will promote the proliferation of nuclear weapons, and how much will the learning cost? There is a perfect analogy here with direct costs. Proponents of nuclear power have argued that it will produce the cheapest electricity, while others have argued that it will be more expensive than fossil generation, and the argument has gone on for years with little sign of movement on either side. As before, the lesson is that nobody knows whether nuclear power will be cheaper than other sources of energy. The question therefore becomes the same as before, how long will it take to learn about the economics of nuclear power, and what will this learning cost? In the case of nuclear power in general, and especially the breeder, learning about direct economic costs or about the technology's effect on the proliferation of nuclear weapons is bound to be slow and expensive because of its great inflexibility. Data, whether about generating costs or proliferation, will accumulate very slowly so that learning will be slow. In addition, because of the capital intensity of the technology, there is little that can be done to control whatever ill effects it produces, whether they arise in the form of a misuse of investment resources or the promotion of nuclear proliferation. Even if nuclear power is found to be more expensive than fossil generation, there is no room for a shift from nuclear power to fossil fuels, the nuclear plant's capital intensity effectively precludes this. The same would be the case if the breeder were found to promote nuclear proliferation. The world-wide abandonment of breeders in an attempt to limit proliferation would be hugely expensive, so much so that some other less radical control of the problem would have to be found.

It can be seen that indirect costs such as the promotion of nuclear proliferation can be treated in exactly the same way as direct costs have been handled. The confinement of the earlier discussion to direct costs is therefore a useful simplification, and not an escape from any special difficulties posed by indirect costs.

(f) *The breeder's declining flexibility*

In chapter 4 a flexible system was seen to be one where errors are quickly discovered (low monitor's response time), where remedies can be applied quickly (low corrective response time), where remedies are inexpensive (low control cost) and where the cost of the system's misbehaviour is low (low error cost). It should be clear from what has gone before that the breeder reactor is a highly inflexible technology. Errors in forecasts of capital costs, load factor, inventory, breeding gain, ORT, reprocessing efficiency, reprocessing costs and fabrication costs all take a long time to be discovered because of the long lead time of the reactor and associated plants, and because of their capital intensity. Large economies of scale favour building them in large units so that data on performance accumulates slowly. Long-range forecasts of uranium prices and electricity demand are also needed, and their time horizon means that a long period is needed to show them to be false. Thus, for errors from each of these sources, the technology has a very long monitor's response time. Whatever errors occur in planning investment in breeder technology are likely to be expensive, giving the technology a very high error cost. Remedies to misinvestment in breeder reactors are all hugely expensive because capital intensive breeders have to be closed down and either replaced by other generating plant, or else electricity demand has to be reduced by substituting non-electrical equipment for electrical. In both cases, the remedy is slow to operate and very costly; breeder technology has a high control cost and long corrective response time. The inflexibility of the technology means that errors in its planning are likely to occur, and whatever errors do happen are likely to be extremely costly.

It is time to stand back from the details of the technology and consider what was needed in order to draw these conclusions. Very little was needed. The breeder reactor has been characterised in only the crudest possible way, nothing being said about even what type was being considered, whether liquid-metal cooled, gas-cooled, with or without carbide fuel and so on. The breeder was taken to be of large unit size, in order to exploit scale economies, of high capital cost and capital intensity and of long lead time, which are all highly general features that breeders of any design will possess. In addition, the reactors' operation requires special reprocessing plant and fuel fabrication plant which, again, were characterised in the broadest possible way. No details about the infrastructure were assumed. But on this tiny fund of information about breeder technology, conclusions of the utmost importance have been drawn which make the technology seem an extremely unattractive investment. The generality of the assumptions made about the technology ensure that these conclusions are generic; they apply to all kinds and varieties and variations of the technology. No breeder reactor system will have anything but the most minimal flexibility, hence the strategy of building breeders must be counted as a highly inflexible strategy. As such it is to be avoided. Breeder technology is not worth further exploration and more R & D in this direction is a mis-use of scarce resources, unless some special case can be made out, a possibility to be discussed, and rejected, in the following chapter.

If strategic decisions about the direction of R & D had been made in the way advocated here, then no resources would even have been spent on R & D for breeder technology. But money has been spent, and considerable technical knowledge has been acquired. Indeed, the breeder has represented a major item in the energy R & D budgets of many countries. There is obviously a feedback between the results of such R & D and estimates of the technology's flexibility made at the strategic level. As more is learned about a technology, judgements of its flexibility are likely to change. If the more that is discovered, the greater the flexibility, then celebration is called for. In the case of the breeder, however, things have gone quite the other way. As more has been learned about nuclear technology in general, and the breeder in particular, the flexibility of the breeder has been diminished even from the pitiful figure it was in happier days. This has happened in three ways as knowledge has been acquired of the general problems of building reactors, of the difficulties in reprocessing spent fuel and of the linear doubling time which can be expected of the breeder. These will be discussed in turn.

As we have seen, the history of thermal reactors has shown that lead times are considerably longer than first expected, and that capital costs are likely to rise continually as more and more safety problems become identified. These features are likely to be found in any kind of breeder technology as well, and they reduce its flexibility. Increases in lead time mean that learning is even slower, or formally, that the monitor's response time is further lengthened. Growing capital costs also increase the cost of misinvestment in breeders, or increase the technology's error cost.

What has been learned about reprocessing is even more depressing for the breeder. In the past, spent thermal fuel was reckoned to need reprocessing in order to make waste treatment and disposal safe. During its time in a thermal reactor some of the uranium in the fuel is transformed into plutonium which reprocessing extracts. Plutonium was therefore seen as a free by-product of the operation of thermal reactors, whose spent fuel would have to be reprocessed for waste management whatever happened to the plutonium so produced. The costs of reprocessing therefore fall entirely to the thermal reactor and not to the breeder, which has merely found a way of using the free plutonium produced from thermal reactors. This view is now beginning to change. Attempts at commercial reprocessing of thermal oxide fuel from LWRs have all failed. In the United States the West Valley plant was closed in 1976 and the Morris plant in 1974. The Eurochemic plant in Belgium was closed after eight years' operation in 1974 and the West German BAK plant has been abandoned after ten years of constant trouble. The French HAO and UP2 plant at La Hague has achieved only one-tenth of its planned capacity. Metal fuel from Magnox reactors continues to be reprocessed, but official estimates of the cost of reprocessing have risen by a factor of ten in the past decade, and are forecast to increase three-fold between 1980 and 1987. Official French estimates for back end costs have increased from Ff 435/kg reprocessed in 1970 to Ff 4,000/kg in 1981 (Finon (1982)). Recent changes in hazard regulation have also made reprocessing look much less attractive for waste management than previously thought. There is therefore a growing interest in once through reactor systems with long term storage of spent fuel (Finon (1982), Sweet (1982) and Schapira (1982)).

The effects of this change on the economics of breeder reactors will be ruinous. If reprocessing is not needed for waste treatment, the costs fall to the breeder programme which needs its plutonium, and costs have risen to such a level that a totally unrealistic uranium price is needed to make the breeder competitive with thermal reactors. Once again, this illustrates the problem of slow learning, which causes so many problems in the planning of nuclear technology. Learning about thermal fuel reprocessing has been extraordinarily slow and in the light of optimistic forecasts which presented plutonium as a free by-product, huge and irrecoverable R & D investments have been made in breeder technology. The change in reprocessing economics affects not only the costs of the breeder, but its flexibility as well. Thermal reprocessing plants would have to be included as part of the infrastructure of breeder technology. This means that there is even greater room than before for mistakes in planning breeder investment, from mistakes in investment in thermal reprocessing plants. At the extreme, planned thermal reprocessing plants may be found to be unworkable, as has happened so often in the past, or they may work but at a cost which makes electricity from breeder reactors more expensive than that generated by other means. The extra infrastructure will also tend to make whatever planning errors are made more costly. For example, if breeder reactors are found to have too high a capital cost to make them competitive with rival fuels, then the reactors themselves will be abandoned once those that have been built are decommissioned, but the capital intensive plants making up the infrastructure will also be abandoned, adding to the cost of error. If these are to include thermal reprocessing plants, then this will add further to the cost of mistakes.

The third problem concerns the linear doubling time of the breeder system, the time it takes for a reactor to produce enough plutonium to fuel a second reactor of the same size, allowing for the plutonium that is inevitably held up in storage, reprocessing, transport and fuel fabrication, and that which is lost because reprocessing cannot be completely efficient. Matters have become steadily worse for the breeder in this respect. In happier days there was talk of doubling times beginning at thirty years, but rapidly falling to between seven and twelve years so that the breeder programme could rapidly substitute for other types of base load plant, producing enough plutonium to fuel the additional capacity needed to meet a growing electricity demand. The growth in demand which can be met from a breeder programme is about equal to the inverse of the linear doubling time. As research on breeder and reprocessing technologies has continued, estimates of doubling time have lengthened to forty, then fifty, then sixty years, until there is now talk of the reactor which does not breed at all, but just manages to produce the amount of plutonium it consumes. Reprocessing of breeder fuel now looks to be more difficult than ever before, and early optimism about breeding gain has retreated to such an extent that the linear doubling times which now seem realistic are much longer than expected originally. As research has gone on, the original inspiring story about the breeder rapidly displacing thermal reactors to give an electricity system independent of fuel supply has become less and less credible.

It is now clear that a programme of breeder reactors would not be able to grow fast enough to meet any but a very modest rate of increase in demand for electricity. If demand increases more rapidly, then the breeder programme will simply run out of plutonium. Nuclear optimists suggest that whenever this happens, more thermal nuclear reactors should be built so that the plutonium they produce may be separated by reprocessing and stockpiled until there is enough to begin building breeders again. This is bound to make breeder technology more expensive because orders for breeders and thermal reactors will be very lumpy and because the thermal reactors which are built from time to time will produce more expensive electricity than breeder reactors would have done had there been enough plutonium. But as before, the flexibility of the breeder is worsened as well as its economics. Thermal reactors and their reprocessing plants would now be part of the infrastructure of breeder technology. They would be needed, not because of their intrinsic worth, but simply as producers of fuel for breeder reactors. They would be needed for this whatever expenses and problems their operation might involve. Since there is now more infrastructure which must be got right if the breeder is to work, then there are more sources of error in planning breeder investment. Thermal reactors may be built when breeders could have been, perhaps because of erroneous forecasts of electricity demand. On the other hand, breeders may be built when there is not enough plutonium to fuel them. The problems of timing investment in nuclear plant have been seen to be very severe, but we are now being asked to switch orders from breeders, with a lead time of between ten and fifteen years, to thermal reactors, with a lead time of between seven and twelve years, and back again as plutonium supplies vary. Moreover, the additional capital intensive infrastructure makes any errors which do happen even more expensive. A thermal plant once built must operate on base load for thirty to thirty-five years, ensuring that if there was enough plutonium to fuel a breeder, the cost of error is high. Similarly, a breeder of even higher capital intensity, which cannot be fuelled because forecasts of plutonium output have been wrong, is an extremely costly misinvestment. Referring back to Table 9.1, it can be seen now why the last two items in the list of direct costs for the breeder are the supply and cost of uranium from thermal reactors.

What has been shown in this chapter is that whatever the details of breeding gain, reprocessing losses and so on, the breeder reactor has always represented a much greater economic risk than ever recognised in the scores of reports to which it has been subject because of its great inflexibility. The decline in expectations about the breeder's performance discussed in this section has worsened the problem by diminishing the small degree of flexibility which it once might have had. But the flexibility was never high enough to make the breeder an attractive technology, certainly not enough to justify its pre-eminent place in so much energy planning. The best breeder reactor would be less flexible than the thermal reactors of today, which have proved to be so troublesome on this score.

3. Conclusion

This chapter has shown how inflexibility may be identified in a technology at a very early stage in its development when it is a candidate for further R & D, at a stage, that is, when very little is known about the details of the technology. This has been shown through the example of the breeder reactor. In its very earliest days it was clear that, whatever the fine points of its design, any breeder reactor would have a high capital cost, a high capital intensity, long lead time, large unit size and would be dependent on special infrastructure for its operation, features which ensure a very low degree of flexibility. Even at this point, therefore, it may be concluded that the breeder is a highly inflexible technology, and errors in its planning will be both likely and expensive. It is, therefore, a most unattractive technology. If the rule of rejecting inflexible strategies defended in this chapter is applied, it follows that there should be no further R & D on breeder technology, unless a special case can be made out for it. What such a case might be is the topic of the following chapter. The breeder's inflexibility would also, of course, prevent its control through the normal political processes of partisan mutual adjustment.

References

Armstrong, M. (1983), *Flexibility in Project Assessment*, Ph.D. thesis, University of Aston, Birmingham.

Buckley, C., G. Mackerron and J. Surrey (1980), 'The International Uranium Market', *Energy Policy*, 8, 84–104.
Chow, B. (1980), 'Comparative Economics of the Breeder and LWR', *Energy Policy*, 8, 293–307.
Collingridge, D. (1982), *Critical Decision Making*, Frances Pinter, London.
Collingridge, D. (1983), 'The Criticism of Preferences', *Philosophy*, 58, forthcoming.
De Neufville, R. (ed.) (1976), *Airport Systems Planning*, Macmillan, London.
Department of Energy (UK) (1975), *Submission to the Royal Commission on Environmental Pollution, Study of Radiological Safety*, London.
Department of Energy (UK) (1983a), *Investment in Energy Use as an Alternative to Investment in Energy Supply*, London.
Department of Energy (1983b), *Transcript of Evidence to Sizewell B Public Inquiry, Snape*, 42, 43, March 17–18, London.
Finon, D. (1982), 'Fast Breeder Reactors—the End of a Myth?', *Energy Policy*, 10, 305–21.
Häfele, W. (1973), *The Fast Reactor as Cornerstone for Future Large Supplies of Energy*, International Institute for Applied Systems Analysis, Laxenburg, Austria.
Keck, O. (1981), *Policy Making in a Nuclear Programme—The Case of the W. German Fast Breeder Reactor*, Lexington, Mass., Lexington Books.
Komanoff, C. (1981), 'Sources of Nuclear Regulatory Requirements', *Nuclear Safety*, 22, 435–48.
Marshall, W. (1981), 'The Peaceful Uses of Plutonium—an Economic Strategy', *Chemistry in Britain*, 17, 466–8.
Marsham, T. (1977), 'The Fast Reactor and the Plutonium Fuel Cycle', *Atom*, 253, 297–311.
Monopolies and Mergers Commission (UK) (1981), *Report on the CEGB*, London.
Nicholson, R. and A. Farmer (1980), 'The Introduction of Fast Breeder Reactors for Energy Supply', *Uranium and Nuclear Energy*, Mining Journal Books, London.
Radetzki, M. (1981), *Uranium—Strategic Source of Energy*, Croom Helm, London.
Richels, R. and J. Plummer (1977), 'Optimal Timing of the US Breeder', *Energy Policy*, 5, 106–12.
Roskill (1971), *Report of the Royal Commission on the Third London Airport*, HMSO, London.
Royal Commission on Environmental Pollution (UK) (1976), *6th Report —Nuclear Power and the Environment*, HMSO, London.
Schapira, J. (1982), 'Some Problems in the Back End of the Nuclear Fuel Cycle', paper to *Issues in the Sizewell B Inquiry Conference*,

Centre for Energy Studies, Polytechnic of the South Bank, London, 26-29 October.

Sweet, C. (1982), 'Logistical and Economic Obstacles to a Fast Reactor Programme', *Energy Policy*, **10**, 15-24.

United States Atomic Energy Commission (USAEC) (1974), *Liquid Metal Fast Breeder Reactor Program: Environmental Impact*, WASH-1535, USEAC, Washington D.C.

United States Energy Research and Development Administration (ERDA) (1975), *Liquid Metal Fast Breeder Reactor Program: Final Impact Statement*, ERDA-1535, ERDA, Washington D.C.

United States National Academy of Sciences (NAS) (1979), *Energy in Transition 1985-2010—The Final Report of the Committee on Nuclear and Alternative Energy Systems*, Freeman, San Francisco.

Vaughan, R. and A. Farmer (1976), 'The Fast Reactor—Energy Without Depletion of Natural Resources', *Proceedings of the Institute of Mechanical Engineers*, **190**, 163-D63.

World Energy Conference (1978), *Nuclear Resources, the Full Report to the Conservation Commission of the World Energy Conference*, IPC, London.

10   SPECIAL PLEADING

The conclusion of chapter 9 was that there is a prima facie case for rejecting the breeder reactor in the strategic choice of R & D programmes for the provision of useful energy in the future. That is to say that R & D on the breeder should not be undertaken unless a special case can be made for investing in a technology of such extraordinarily low flexibility. This chapter will therefore concern itself with two arguments which attempt to show that the breeder is a good technology despite its inflexibility. The two arguments are often heard about other technologies besides the breeder, and so the discussion should have something to say beyond the particular example of the breeder reactor. The first argument is that the breeder is *inevitable*, the second that it is needed as a *hedge* against the worst possible energy future which might happen.

1. The breeder is inevitable

Proponents of breeder technology have argued in the past that it is inevitable because all natural resources are limited. Continuing use of coal, oil and natural gas will, it is argued, eventually mean that the easily exploited deposits are exhausted, so that future supplies will have to come from reserves which are much more expensive to work, and that the prices of these fuels will eventually reach very high levels and remain there. The same is true of uranium: sooner or later thermal reactors around the world will have consumed so much of the fuel that its price will go up in the same permanent way. Since the low cost reserves of all these fuels are limited, there must come a time when they all reach a high price, after which energy supply will become increasingly difficult. Here, of course, the breeder steps in. Since the breeder employs plutonium from spent thermal reactor fuel for which there is no other use, and can also convert the otherwise valueless stocks of depleted uranium into plutonium which can fuel further breeder reactors, it can produce energy without the depletion of natural fuels. Breeder reactors will have higher capital costs than thermal reactors, and much higher ones than fossil generating plant, but the very high price of fossil fuels and of uranium will ensure that the breeder will be able to produce electricity at a lower cost.

Because the breeder is inevitable, in the sense that there is certain to come a time when it provides the cheapest electricity, then even if it is very inflexible it should still be developed. Indeed many of the problems caused by its inflexibility are moderated if the technology is once admitted to be inevitable. The problem of timing investment in a breeder programme discussed in the latter part of the previous chapter, for example, is lessened if the technology is seen as being needed sooner or later. Too early an investment merely means that experience is acquired a decade or two before it is needed, but at least it is going to be needed at some time. The fear of investing in breeder technology only to find that it is never economic is banished.

Several studies have attempted to quantify the argument. OECD (1979), for example, looks at the likely rate of construction of nuclear reactors throughout the western world and the consumption of uranium that this would imply, with varying assumptions about the types of reactor available and their market penetration. The study accepted the projections of nuclear power capacity throughout the western world which were made by the International Nuclear Fuel Cycle Evaluation. Various scenarios about reactor type and penetration of the overall market for reactors were then constructed, and uranium demand calculated for each. This could then be compared with the expected supply of uranium from conventional low cost resources. Figure 10.1 shows that even for the low growth projection of nuclear capacity, uranium supplies are exceeded before 2000 if LWRs are used with no reprocessing of their spent fuel——the current technology once through LWR strategy. The upper mixed strategy allows 10 per cent of nuclear capacity to be from breeder reactors from 2025, while the lower mixed strategy takes this proportion to be 65 per cent. In the large-scale advanced fast breeder reactor strategy, advanced carbide-fuelled breeders capture the entire American and West European markets for reactors. The only constraint is that their penetration may be limited to as low as 85 per cent of the total reactor market by the world-wide supply of plutonium, the remainder

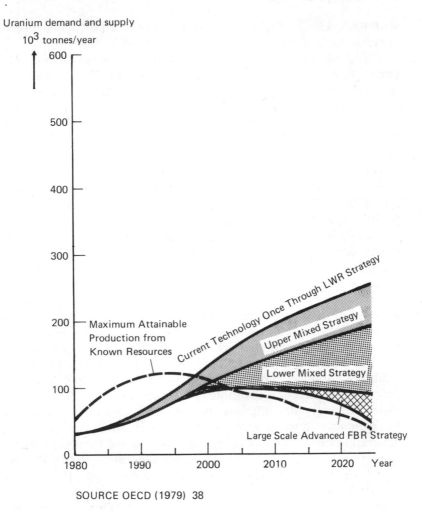

Fig. 10.1. Comparison of annual world uranium supply and demand to 2025 low growth projection.

being made up of LWRs. It can be seen that the other reactor strategies also lead to a greater demand for uranium than can be supplied from the reasonably assured resources and estimated additional resources of known uranium deposits. All strategies therefore lead to the need to find new resources of uranium, but the large-scale advanced fast breeder reactor strategy obviously causes fewer problems here than its rival strategies. The problem is, of course, worsened if higher projections of nuclear capacity are made.

The study does not pretend to be forecasting nuclear energy capacity, nor what reactor strategies will actually be followed, nor what uranium supplies will be available; rather it is seeking to explore the size of the problem caused by limited uranium supplies. It claims that even with modest assumptions about the rate of growth of nuclear capacity, the strategy which is most conservative of uranium still leads to problems of supply in the future. This is worrying, since nuclear power may grow at greater rates than assumed, and much less conservative strategies might be followed. It looks, in other words, as though the limited nature of uranium resources makes the breeder reactor inevitable well before 2020. Other studies have come to similar conclusions, notably the World Energy Conference (1978) and the United States National Academy of Sciences (1979).

These studies are, however, already looking a little sorry for themselves. They all make a number of very questionable assumptions. It is assumed that electricity demand will rise at what looks at the moment to be quite unrealistically high rates over the study period. This has been discussed already for the United States in chapter 1, which considered a similar study claiming to show the need for an intensive R & D programme for breeder technology in that country. The second assumption is that nuclear power will have a marked and unambiguous cost advantage over fossil fuels for base load generation throughout the world. This, as we have seen in the historical discussion in Part II, cannot be claimed on the evidence of past investment in nuclear technology. The rate of substitution of nuclear for fossil power has been far lower than that assumed in these studies. The World Energy Conference (1978) takes installed nuclear power in the western world in 2000 to be 1,540 GW, and its 1985 forecast is 303 GW. As Buckley *et al.* (1980) show, the 1985 figure cannot be as much as 220 GW and capacity for 2000 is hardly likely to exceed 700 GW. The OECD (1979) puts capacity in that year at between 800 and 1,250 GW, while the United States National Academy of Sciences (1979) gives 250 to 500 GW for the United States alone. Fossil prices may not, after all, increase to the extent needed to give a definite cost advantage to nuclear power.

The third assumption concerns the possibility of reprocessing spent LWR fuel, which we have seen to be essential for the extraction of plutonium to fuel breeder reactors. The problems which have arisen here have been discussed in the previous chapter. No commercial oxide reprocessing plant has worked anything like as efficiently as planned and many have been scrapped. Costs have risen by a factor of about ten in the past decade, which makes breeder technology much less attractive. If reprocessing is needed only to extract plutonium for breeders, then breeders must bear its cost, which would make them much more expensive than these studies assume, even if large-scale reprocessing becomes technically possible. The final assumption made in the studies is that future discoveries of uranium resources will be limited, so that the price will rise to levels favouring the breeder. Pessimists like the United States National Academy of Sciences (1979) point to the poor results of surveying for new resources in the United States. Optimists, like Buckley *et al.*, and Radetzki (1981) argue that extreme underestimates of reserves are reasonably universal for all minerals, being a function of industry's reluctance to invest in exploration once reserves with a lifetime of between fifteen and twenty years are available. The long period of depressed uranium prices until 1973 did nothing to encourage further exploration, and much of Africa, Central America and Southeast Asia have gone unprospected. Like those for all minerals, resource estimates have steadily increased as exploration has continued and surveying techniques have improved, and better mining and processing technologies are likely to bring in many additions to the reserves, which were previously too expensive to be exploited. Thus, the exhaustion of low cost reserves of uranium within the next few decades is far from being a certainty.

In short, studies which seem to indicate that the breeder reactor is inevitable as a response to very high fossil fuel prices and high uranium prices which are bound to be found sooner or later, make a number of dubious assumptions to reach that conclusion. The picture they paint of the future may be correct, but this is not the point. It may or may not be correct, but it is not inevitable. The breeder reactor is not an inevitable technology, and this attempt at special pleading fails. We may now turn to the second attempt to make out a special case for the breeder.

2. The breeder as a hedge

The second special case for continuing efforts to achieve breeder technology despite its great inflexibility is that it is needed to guard against the worst developments in energy supplies. The worst that may happen is the existence of permanently high prices for fossil fuels and for uranium. It is admitted that this is not inevitable, there may never come a time when the breeder is economic, but early development of the breeder is needed in case the worst happens. In the worst state of energy supplies, the breeder would be a valuable and economic source of energy, but the lead time for its development is so long that it needs to be developed now and cannot wait until the worst appears to be coming about. In the words of Nicholson and Farmer (1979):

The full exercise of the fast reactor option will take some time because of the learning processes and the investment involved. At the time it is exercised, pressure on energy resources may be great. If the fast reactor is to be most effective in meeting energy demands, it is necessary to complete the demonstration and development phases in good time. The option will be the more valuable the more fully it has been demonstrated and therefore the more quickly it can be exercised when the need comes. Lead-times are long and, in the authors' view, the likelihood that fast reactor introduction will be a gradual process is no argument for delay in current development programmes.

The breeder is here proposed as a hedge, in the sense of an option chosen to avoid the worst that might happen, in this case high fossil and uranium prices without breeder reactors. 'Hedging' can also mean spreading bets over a number of out-comes in a gamble, but this is not the sense in which it will be used here. In making decisions under extreme uncertainty, or ignorance as I prefer to call it, the strategy of hedging is often recommended. It has great appeal, not only for the breeder reactor, and so it is worth examining in some detail and at some level of generality. I hope to show, following Collingridge (1983), that hedging is never a reasonable strategy, not for the breeder nor in any other case. Once the general problems inherent in hedging have been examined, those which are peculiar to the breeder will be discussed. Even if hedging were a reasonable strategy for coping with extreme uncertainty, the breeder has features which make it extraordinarily clumsy in this role. The strategy of hedging calls for the worst which might happen to be avoided. The decision maker is to compare the worst outcome which might occur for each of the options which is open, and is then to choose the option with the least worst outcome. If the decision problem is sufficiently well structured, the strategy of hedging is equivalent to the maximin rule. In the real world, however, such choices are rare, and so it is best to give hedging a somewhat wider scope than the formal maximin rule. For example, in many choices the worst possible outcome is not well defined, so that maximin cannot be applied strictly even though it is still useful to talk of hedging in the sense of avoiding very bad outcomes. Hedging is therefore a counsel of caution; it seeks to limit, and in well structured decisions actually to minimise the damage if the worst happens. Herein lies whatever plausibility the strategy may have because caution is obviously appropriate under conditions of such extreme uncertainty. I hope to show, however, that whatever its plausibility, hedging is not a rational strategy for choosing under conditions of ignorance. I shall argue that the strategy leads to several intractable problems, each of which will be illustrated by an example of a decision about technological change. In its place, I propose a strategy of *flexing*. The name of this strategy derives from 'flexibility'.

The central idea here is the realisation that no decision made under conditions of ignorance can be known to be correct; that is, there may always be an option other than the one taken which the decision maker would regard as preferable to the original if he were aware of it, or aware of additional information which was not possessed when the original decision was made. It follows that any decision under ignorance may turn out to have been mistaken or erroneous. If the decision maker becomes aware of a superior option which was previously overlooked, or if facts which he discovers make him revise his original preferences for the chosen option over all others, then the first decision may be said to be erroneous. This is not, of course, to cast aspersions on the skill of the decision maker; he may well have been doing his best to solve a very difficult decision problem. Since any decision under ignorance may prove to be erroneous, a rational agent should positively look for error—should, that is, *monitor* his decision. He should continue to search for options which have been overlooked which might prove preferable to the option originally taken, and should look for facts which show his original choice to be inferior to some other option. But the discovery of error is pointless unless something can be done by way of correction, and so the rational decision maker should favour options which can be revised if they are found to be in error.

A decision maker thus armed with the resolve to make flexible decisions and to monitor them can afford to take risks; there is no longer any need for the extreme caution involved in hedging. Consider the simplified problem of Figure 10.2, where there are two options, A and B, each with two possible outcomes; A may lead to A1 or to A2, and B may lead to B1 or to B2. Let A1 be better than A2, and B1 better than B2, B2 better than A2, and A1 better than B1. This is a very common kind of decision problem because A has the best of all outcomes, A1, but also the worst of all of them, A2. Hedging, here equivalent to maximin because the problem is artificially structured, tells the decision maker to choose B. The worst that can then happen is B2, but at least this is better than choosing A and obtaining the outcome A2. Hedging tells the decision maker to forego the best so as to avoid the worst.

Flexing, on the other hand, tells him to try for the best, and to risk the worst, but also to be ready to reverse his decision if the worst happens. A would be chosen on the strategy of flexing, but once the decision had been taken it would be monitored in case A leads to A2 and not the hoped for A1. At the same time, the decision maker should take steps to ensure that the choice of A could be easily reversed if it is found to lead to A2. This generally involves acting so as to reduce the time needed to change from A to B if A2 occurs (reducing the *response time*). Since the decision can be reversed without too great a cost, if A2 is found to occur, the worst may be avoided and B substituted for A, so that the final outcome is either B1 or B2, both of which are better than A2. Because A2 can be avoided in this way, the decision maker can take a risk and try to arrive at A1, the best of all the outcomes. This will become much clearer when the two strategies are applied to examples below. Each example illustrates some deficiency in hedging which does not arise in flexing.

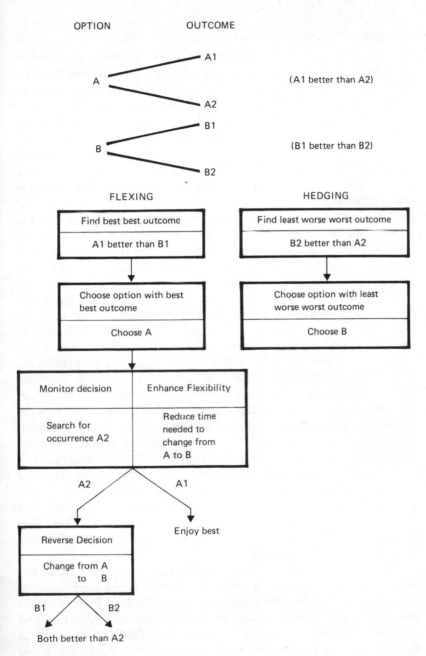

Fig. 10.2. Hedging and flexing.

(a) *Hedges can rarely be revised*

The first problem about hedging as a way of living with ignorance is that it fails to consider whether or not the decision to hedge can be revised if the feared worst outcome does not happen after all. There may be something to be said for hedging when the hedge can be revised if the worst does not happen, but where revision is impossible hedging must be much less attractive. A good example here is the decision to develop the first British nuclear energy programme based on the gas-cooled Magnox reactor, discussed in chapter 5. This decision was essentially a hedge against the continuation of the coal shortage which had existed for several years. The case for the programme was made in the 1955 White Paper, which argued that the country's growing economy might require a considerable increase in energy use, especially electricity. The size of the generating system which would be needed to cope with this increase in demand was put at 55 to 60 GW by 1975, compared to the 20 GW of 1954. If coal was to be used for power generation on this scale, the quantity burned in power stations would have to increase from 40 million tons to 100 million tons per year by 1975. It was thought very unlikely that the coal industry would be able to meet such demand. The White Paper therefore proposed a rapid development of nuclear power, the early part of which would not be economically competitive with coal burning generating stations, although it was expected that later reactors would be competitive as they would benefit from learning. In March 1957 the Government announced that the original programme was to be expanded to between 5 and 6 GW of nuclear power by 1965. The Suez crisis of the previous October had cast doubts over the future availability of oil; and even if this fuel was available its importation would place a great burden on the British balance of payments.

The decisions to build the Magnox programme were therefore largely a hedge against the British economy not being able to grow because of the shortage of coal and oil, at least without placing an intolerable strain on the balance of payments. But by the time the programme was even half finished it was clear that the feared shortage of coal and high price of oil were not going to happen. By 1959 the National Coal Board had stocks of 40 million tons of coal. Oil was also readily available at much lower prices than previously expected, and there was a glut of oil tankers. But by this time seven Magnox stations had been ordered or were under construction, so there was very little room to alter the programme. It was marginally reduced in size, and spread out over a few more years with a planned nuclear capacity of 5GW by 1968. As a hedge, the Magnox programme was not needed but by the time this was known there was very little scope for reversing the programme.

What would have happened if the strategy of flexing had been applied instead? Figure 10.3 is a simplified outline of flexing in the Magnox choice. The order of preference of the outcomes is taken as energy balance with no nuclear power, energy balance with nuclear power, energy surplus and energy shortage. If a balance between the supply and demand for energy could have been achieved without the expensive new technology of nuclear power, this would have been better than relying on nuclear power for its achievement. An energy surplus indicates an expensive misinvestment of resources making it inferior to any outcome with a balance between supply and demand, but it is clearly preferable to bearing the economic losses of an energy shortage. The option leading to the best outcome may lead to the worst, so hedging would favour the other option, a Magnox programme, sacrificing the best outcome to avoid the worst. Flexing, on the other hand, would have led to the refusal to develop a Magnox programme, since the best of all outcomes is achieving an energy balance without nuclear power. This is, however, to risk an energy shortage and the proponents of hedging argued that if the nuclear energy programme was started only when an energy shortage could be foreseen with a very high probability, there would be many years of costly energy shortage. But flexing requires action to reduce the cost of the worst if it happens. In this case action must be taken so that any impending energy shortage will be detected quickly and will be able to be remedied quickly without great damage to the economy. Steps would be taken to ensure that any gap between supply and demand could be closed quickly ensuring that energy shortages do not constrict the economy. There are innumerable ways in which this might be done: by programmes to improve the

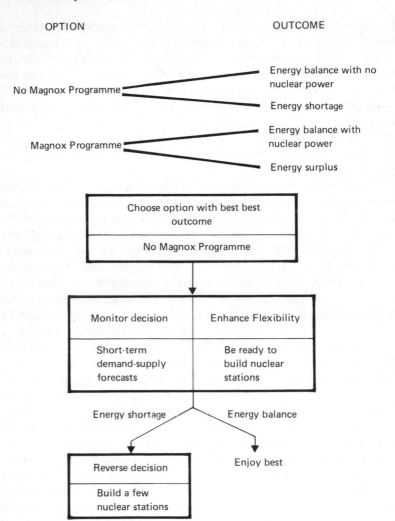

Fig. 10.3. Flexing in Magnox decision.

efficiency with which energy is used; by searching for renewable sources of energy which can be developed quickly if needed or by developing nuclear reactors like Magnox to a stage where they can be brought into operation very quickly. For the sake of simplicity Figure 10.3 considers just the last of these. If monitoring was then to reveal an impending gap between supply and demand, nuclear stations could have been built rapidly to remedy it without damage to the economy. The feasibility of this approach will be examined a little later.

In no way should the adoption of flexing be seen as a prescription for inaction. Flexing here calls for action to reduce the costs of an energy shortage should it occur, for example, through a very extensive R & D programme on nuclear power, perhaps with the construction of demonstration reactors, or an intensive search for other, perhaps renewable, energy sources. Flexing gives the decision maker the chance of achieving the best outcome—it is in no way a policy of do nothing and hope for the best. Nor does flexing unfairly favour the status quo. Even the most convinced advocates of nuclear power must admit that the technology should only be developed when the time is right and historical episodes like the Magnox programme show how expensive mistiming can be. The lessons to be drawn from this are that hedging tends to lead to the premature adoption of technologies. This may be avoided by a strategy of flexing, which ensures that a technology is developed only when it is needed. Flexing is not a decision strategy in opposition to technological change; it is intended to enhance the benefits from such change by improving its timing.

Another way of looking at the matter is that flexing improves the flexibility which the decision maker has by increasing the number of options which are open, unlike hedging which closes options. Once the Magnox programme was underway as a hedge against future energy shortages there was very little flexibility; the final size of the programme might be varied a little up or down but it was soon too late to abandon it altogether. By the time it was realised that there was not going to be the energy shortage which was so feared, the option of not having a Magnox programme had closed and only marginal changes could be made in the original plan. Ordering a whole number of stations together had severely restricted the ways in which the new technology could be developed, so much so that the ability to react to new information about the availability of fossil fuels was lost. If flexing had been employed, on the other hand, the option to have no nuclear programme would have been retained and with it the ability to respond to information as it became available. If monitoring showed no impending energy gap and abundant supplies of fossil fuel, then the nuclear programme could have been delayed further, although the R & D associated with it would have continued. If monitoring was one day to reveal an impending gap between supply and demand, then the nuclear programme could be started in the knowledge that it was really needed. The effect of the R & D is to increase the options for the new technology's development because it can be deployed more quickly. Thus where hedging destroys options and restricts the decision maker's freedom and ability to respond to new information, flexing opens up the future, promotes flexibility and eases the adaptation of policy to information.

(b) Hedging often brings about the worst outcomes

Many decisions which have to be made under ignorance are ones involving competition with another decision maker; indeed it is often the inability to forecast what choices a rival decision maker will make in the future which creates the state of ignorance. Under such circumstances it is well known that hedging by one decision maker often brings about the very outcome which is feared. An example here is the American decision to build a hydrogen bomb discussed by York (1975, 1976). An American H-bomb was proposed following the first Soviet A-bomb test in 1949. The idea was examined in depth by the General Advisory Committee to the recently formed Atomic Energy Agency. The Committee were unanimously opposed to the programme, employing a whole battery of forceful arguments against it. Nevertheless, the project began to gather supporters whose principal argument was that an American H-bomb was needed as a hedge against the Soviet Union acquiring such a weapon, whose feasibility was by now established. The new weapon's supporters won the argument and in March 1950 President Truman announced that a crash programme would be started. The first proper H-bomb was tested in spring 1954. The Soviet Union soon learned of American intentions. There is no way of knowing whether the Soviet Union had plans for an H-bomb before this, but what is sure is that the American programme gave the Soviet Union a sufficient reason for acquiring the new weapon. The first Soviet test came in November 1955, and so both sides had the new weapon.

These self-fulfilling predictions are well known to defence

planners, but they argue that it is better to hedge, even if both sides end up with a new weapon when neither really wanted it, than to risk a long period when only one side has a new weapon. Thus, even if the Soviet bomb was developed purely as a counter to the American one, both sides having the new bomb is still better than the risks which would have followed only one side having it. Figure 10.4 illustrates how the decision might have been handled with flexing. Since the best of all the outcomes is both sides having no H-bomb, flexing indicates that the American bomb should not be constructed. This, however, risks the worst outcome, where the Soviet Union but not the United States has the bomb. Flexing calls for monitoring, in this case searching for signs of a Soviet H-bomb programme or tests, and for the costs of the worst outcome to be reduced by ensuring that the original decision is easily reversed. The United States could have developed the new weapon to the stage of its first test, so that at the first sign of a Soviet H-bomb test American weapons could have been deployed within a very few years. In this way, there need be no danger from long term imbalance in the deterrent forces of the two sides. Indeed, this was urged in the General Advisory Committee's report. Herbert York offers further analysis and shows convincingly that with appropriate R & D the United States would have been able to deploy its H-bomb with sufficient speed for there to have been no risk from waiting for firm knowledge of a Soviet bomb, rather than hedging against a speculated Soviet bomb.

This is an example where the opportunity of flexing was clearly open; the United States would have risked no significant reduction in its security by such a policy. Even if the Soviet Union tried to take advantage of American flexing by developing its own H-bomb as quickly as possible, the United States was sure to detect their first tests and, given the intensive R & D programme demanded by flexing, would have been able to develop a counter weapon before the balance of terror could be disturbed. In fact, flexing has been employed in strategic decision making, for instance in President Carter's policy on the neutron bomb.

As in the Magnox example, flexing improves the timing of a decision to adopt a technology, ensuring that it is developed only when needed. Hedging, as before, may have led to a

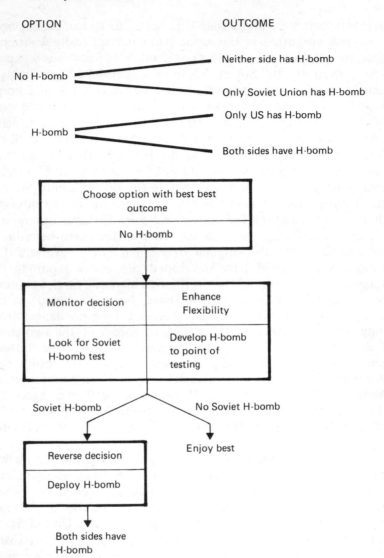

Fig. 10.4. Flexing in American H-bomb decision.

premature adoption of the H-bomb. The H-bomb may have been inevitable, given the policy of deterrence adopted by both powers, but this is not to say that its timing is of no importance. A premature H-bomb must be seen as very costly in terms of unnecessarily increasing the pace and investment demands of the arms race. Flexing would have countered this by ensuring that the United States developed the bomb only when it was needed to ensure its security. In this case too the strategy of flexing can be seen as promoting the decision maker's flexibility and his ability to respond to information.

(c) The 'worst outcome' is often ambiguous

The third criticism of hedging is that in many decision problems the 'worst outcome' is ambiguous, in the sense that hedging fails to point to a single option but instead indicates a number of options. A good example here is the United States' decision to deploy their new technology of multiple independently targetable re-entry vehicles (MIRV) (discussed by Greenwood (1975), Tammen (1973), and York (1975, 1976)). Early rockets carried only one warhead, but MIRV enabled one rocket to deliver a number of warheads at widely separated targets. A complex system like MIRV takes seven to ten years to develop and so forecasts of what weapons the Soviet Union might possess this far ahead must be made. In the early 1960s American strategists were worried that within this time scale the Soviet Union might acquire anti ballistic-missile systems (ABM) able to shoot down incoming warheads. To simplify history, MIRV was proposed as a hedge against Soviet ABM. MIRV enabled each rocket to deliver a number of warheads, overwhelming the radars of the Soviet ABM so that a good many warheads were sure to find their target. The development of the new weapon went ahead very smoothly, and in great secrecy. When some of this secrecy was lifted opponents of MIRV, in particular the Arms Control and Disarmament Agency, objected that it actually made nuclear war more likely and not less. They argued that the Soviet Union would have to place the worst possible interpretation on American deployment of MIRV, seeing it as a counterforce weapon for use in a first strike against the Soviet Union. This, it was held, would seriously unsettle the balance of deterrence, even promoting Soviet intentions to launch a first strike against the United States. As a hedge against this worst outcome, the United States should abandon MIRV. This, and other arguments were not persuasive, however, and MIRV's deployment began in 1970.

For our purposes, the story casts grave doubts on hedging as a strategy for coping with ignorance. Hedging led to contradictory prescriptions about MIRV. Hedging based on the assumptions made by MIRV's proponents about Soviet ABM, calls for the introduction of MIRV. MIRV's opponents, however, suggest that the new technology could lead to nuclear war because of the destabilising perception of MIRV by the Soviet Union as an American first strike weapon. Hedging against this obviously calls for the abandonment of MIRV. The lesson is that without facts to guide the decision makers, fantasies of all kinds fill the void, and action is actually recommended on the basis of fantasy. Proponents of MIRV pick their fantasy and cling to MIRV as a hedge against it; opponents design their own fantasy and recommend MIRV's abandonment.

Does flexing escape these problems? In the case of MIRV flexing indicates one option. The best outcome is for the United States to have no MIRV and the Soviet Union no ABM, so the first choice is not to develop MIRV. Monitoring should search for signs of Soviet ABM and MIRV should be deployed only when positive evidence of this is found. In this way the best outcome can be aimed at and MIRV deployed only when hard facts, and not fantasies, show that it is needed. The standard objection to this, as before, is that delay is too risky; by the time the United States could deploy enough rockets with MIRV the Soviet Union could have enough ABMs to risk an attack. But flexing requires action to reduce the damage which would be inflicted by the worst outcome, should it happen. If MIRV had been developed to the stage of its first test flight, it could have been deployed within a year or two of firm evidence for Soviet ABMs being acquired. This would have been in ample time to counter the danger of an unbalanced deterrent (Collingridge (1980)). Flexing would have enabled decision makers to respond to genuine information about weapons under development in the Soviet Union. Hedging, as before, destroyed this ability, and in its absence fantasy took the place of information.

(d) Hedging can lead to a vicious circle

A great deal of British energy policy can be seen as hedging against a future shortage of energy which would restrict growth in GDP. One example we have already seen is the Magnox decisions. It is not claimed that there is a causal relationship between GDP and energy consumption, nor that there will be a shortage of some form of energy, merely that there may be such a relationship and such a shortage. If this happens, the consequences will be a constraint on economic growth for a number of years until it can be corrected, which would, in a competitive world, be hugely expensive to the British economy. As a hedge against this, immediate action is generally urged to increase future energy supplies.

The present debate about energy futures throws the whole issue into sharp relief. The conventional way of forecasting energy demand is through the projection of past relationships between GDP and energy demand (Department of Energy (1978) and Energy Policy (1978)). Recently, however, a much more disaggregated approach has been adopted by some forecasters such as Chateau and Lapillone (1978), Cheshire and Surrey (1978), Leach *et al.* (1979), Lewis (1979) and Lovins (1979). They look at the end use of energy in a very large category of industries, services and home conditions, paying particular attention to conservation techniques which are already spreading into use, or are on the near horizon, which will improve the efficiency with which energy is used. Their analysis shows that even the large increases in GDP allowed for in conventional planning (a factor of three from 1975 to 2025 for example) can be largely achieved by improving end use efficiency. Very little extra energy is needed in the foreseeable future. The policy implications of such a view are, therefore, very profound.

The official government view rejects the usefulness of such low energy forecasts. Aiming at such a future is just too risky. If people do not conserve energy in the way foreseen by the low energy forecasters, aiming for a low energy future will be hugely expensive because there will be an undersupply with a consequent restriction on economic growth. As a hedge against this, plans must now be made, and indeed are being made, for a massive increase in the provision of energy. The case is succinctly put by Norman Lamont, Parliamentary Undersecretary in the department of energy, who is answering W. Patterson's defence of the low energy forecasters:

. . . if we are to have no nuclear power I would like to know how it is that Mr Patterson assumes that we will be able to satisfy our energy needs if . . . his view of the future turns out to be wrong. What happens if the arguments which he's put forward just don't work out; how are we going to provide the energy needs of this country? [quoted in Collingridge (1980), 100–1]

Energy planning in the past has been conducted in the way favoured by Lamont. Much has therefore been learned about the best ways to increase energy supply, but next to nothing about how consumers will improve the efficiency with which they use energy, and ways of encouraging them in this. Lamont is therefore perfectly right, aiming for a low energy future will be very risky, but only because of the hedging which has dominated energy planning in the past. If Lamont wins the day and hedging leads to a future increase in energy supplies, exactly the same dilemma will arise in a few years' time when the next round of energy planning is due. It will still be possible to try for a low-energy future, but the risks will now be even more than before because energy consumers will have adjusted to plentiful supplies of cheap energy and because we will remain as ignorant as ever about conservation. Hedging will therefore be even more favoured than it is today and energy supplies will have to be increased yet again. The whole story can be repeated over and over again, of course, until it becomes impossible to supply the quantities of energy which are thought necessary or until consumers become unable to adjust to such generous supplies, with a resulting very costly oversupply of energy. I have referred to this situation as a *hedging circle* (Collingridge (1980), chapter 5).

Flexing enables this circle to be broken. The best outcome, the future painted by the low-energy forecasters, should be aimed at, and so there should not yet be any large increases in future energy supplies, such as the government's present nuclear power programme. But this is to risk the worst outcome, a shortage of energy and restriction of economic growth. Flexing therefore calls for monitoring in the form of short-term forecasts about supply and demand of energy. If it is found that consumers are not adjusting as the low-energy forecasters suppose, and are not using energy efficiently, then energy supplies can be gradually increased. Those who favour hedging, of course, argue that this will be very expensive because increasing energy supplies takes a very long time, during which the economy will be constrained. But flexing requires actions to reduce the cost and to reduce the lead time of increasing supplies, such as keeping in the ground readily available open cast coal, and being ready to build smaller power stations more quickly than large ones. Such action will obviously be expensive in the present state of the art of power plant construction but such is the cost of living with ignorance.

(e) Hedging pays no regard to benefits

A common criticism of the maximin rule is that it pays no attention to the benefits which might be obtained from a course of action. All that is relevant to the choice is the worst which might happen for each option; there is no way of balancing this against the benefits which would accrue if a happier state of the world is realised. This can be seen in the previous discussion of low and high energy futures. Low-energy futures are dismissed simply on the grounds that if consumers do not behave as expected, economic growth may be curtailed. Nothing is said about the benefits of a low-energy future, which are very considerable. Table 10.1 compares demand forecasts from the British Government and from Leach.

Table 10.1. Forecasts for Primary Energy Demand 2000

Fuel	Energy Policy (1978) High case (Billion therms)	Leach (1979) High Case (Billion therms)
Solid fuels	17.0	14.5
Gas	17.8	11.9
Liquid fuels	26.5	22.8
Electricity	15.8	8.0
Total	77.1	57.2

The consequences for energy policy are profound. For example: by 2000 the United Kingdom could be self-sufficient in North Sea oil and gas even on central estimates of reserves; coal production need only be about 120 million tonnes per year, well below the 170 million tonnes required on the basis of traditional forecasting; only about 30 GW of electricity capacity need be built until the end of the century, compared with 83 GW on traditional forecasts, with a capital saving of about £30,000 million; electricity demand can be met by building only 6 GW of nuclear capacity in the first quarter of the next century making the nuclear energy question peripheral and fast breeder reactors can be shelved indefinitely. Nothing is said about the huge benefits of such a future. The strategy of hedging dismisses these as irrelevant. As far as official British policy is concerned, a low energy future is simply not a political option which is up for discussion; for all its benefits, hedging demands that it is a road not to be taken. It is clear that the strategy of flexing does not suffer from the same objection. Whereas hedging foregoes the best to avoid the worst, flexing positively seeks the best. Nor can flexing be accused of the converse, that it seeks the best but ignores the worst. Flexing insists that insurance is taken out, in terms of monitoring and enhancing flexibility; if the worst happens, the damage it does is limited.

3. Hedging and flexing with the breeder

I have tried to show that the special pleading for breeder technology which defends it as a hedge against the worst possible future, where there are high prices for coal, oil and uranium, is mistaken because hedging in this way is simply not a reasonable decision rule. But returning now to the particular case of the breeder, there are problems which make it a technology peculiarly unsuited to the hedge for which it is envisaged.

The first problem is that electricity provides only a small fraction of any country's final energy use, thus hedging by building breeder reactors will only safeguard electricity supply, and base load supply at that, ensuring that users of non-base load electricity and non-electric fuels will still have to bear very high energy costs. If the breeder is to offer more than this modest protection, energy users must be persuaded to turn to electricity. The logic is very similar to the electrification policy adopted in France, and discussed in chapter 7, where consumers were encouraged to substitute electricity for other fuels so that there would be the assured demand needed by the PWR programme. There are three problems associated with electrification for the breeder. First of all, it is expensive since consumers are persuaded to switch from cheaper fuels to electricity. Secondly, because replacement of capital equipment is involved the substitution must be gradual, and the security offered by the breeder is slow to develop. Finally, it is incompatible with imposing conservation to ensure efficient use of energy, a policy which can be seen as an alternative way of guarding against high primary fuel prices. If it seems that prices for coal, oil and uranium might all be very high in the medium future, then it might be thought that two ways could be tried to hedge against this unhappy event, building breeders and conservation. But the former, if it is to be at all effective, demands the electrification of the economy, which contradicts attempts to conserve energy. Breeders can be brought in on a large scale and rapidly, as the French have so convincingly shown with their thermal programme, only if there is a guarantee of growing electricity demand, a requirement which cannot be permitted to be undermined by conservation policies.

The second difficulty in proposing the breeder for a hedge against high primary fuel prices is that it takes many decades before breeders can supply a reasonable fraction of base load electricity, a problem which has worsened as the linear doubling time of the breeder has steadily increased as more research has been done, a point discussed in the previous chapter. A policy of conservation would be operative in a much shorter time, but as we have seen, the two approaches to impending high primary fuel prices are contradictory. The final problem in hedging with the breeder is that it actually worsens the security of energy supply by making a large fraction of a country's primary energy depend upon very sophisticated technology, including, of course, its equally sophisticated infrastructure. The cost of failure from any cause would be enormous. The failure might be technical, perhaps with the reactors themselves, or with reprocessing chemistry, or might result from industrial action, the discovery of unexpected safety problems or health hazards, or even the air strikes discussed by the Royal Commission on Environmental Pollution (1976). The whole issue of the vulnerability of electricity supplies has been discussed by Patterson (1977).

All in all, then, the breeder reactor cannot sensibly be suggested as a hedge against high primary fuel prices in the future.

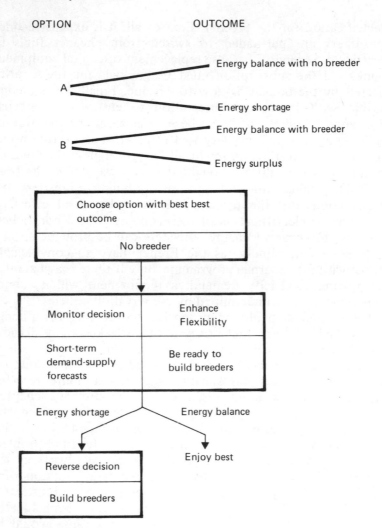

Fig. 10.5. Flexing in breeder design.

But can the strategy of flexing be applied to the breeder? If it can then there is at least that much of a case for continuing R & D on breeder technology. Figure 10.5 shows what this would look like. It is, not surprisingly, very similar to Figure 10.3 concerning the Magnox reactors. The best option is an energy balance with no breeder reactors, and this should be aimed at. The decision not to build a full-scale breeder programme would be monitored by using short-term forecasts about energy supply and demand, and R & D on the technology would be continued in case it was needed. If there is no impending energy shortage, then the best outcome is enjoyed and there is an energy balance without breeder reactors. If, on the other hand, there appears to be an imminent shortage of energy, then the breeder should be brought in as rapidly as possible. The problem concerns the final step. We have noted that the breeder is simply not suitable for a crash programme because with linear doubling times of between fifty and sixty years the programme can grow only very slowly. Thus the substitution of cheap electricity from breeders for expensive electricity generated from thermal reactors and fossil fuel will be spread over many decades. The whole point of the strategy of flexing, however, is to delay choice of the second-best option until there is clear evidence that it is needed, and that the option chosen earlier is leading to the worst outcome, not the best. For breeders, this warning time would have to be several decades, even if R & D were conducted up to its full commercialisation. This is quite impossible, as forecasts of energy supply, demand and costs over such a period are just guesswork. Breeder technology is quite unsuited to the strategy of flexing.

In conclusion, the technology of the breeder reactor is not suited for hedging, even if this were a rational decision strategy, nor is it suitable for the more reasonable strategy of flexing. It seems to be a technology to be avoided altogether because the second case of special pleading fails as much as the first. In the absence of successful counter arguments, the great inflexibility of the technology entails that no R & D programme for the breeder should be funded; it simply cannot be a solution to any problem of energy policy.

References

Buckley, G., G. MacKerron and J. Surrey (1980), 'The International Uranium Market', *Energy Policy*, 8, 84–104.

Burn, D. (1978), *Nuclear Power and the Energy Crisis: Politics and the Atomic Industry*, Macmillan, London.

Chateau, B. and B. Lapillone (1978), 'Long Term Energy Demand Forecasting—A New Approach', *Energy Policy*, 6, 140-65.

Cheshire, J. and A. Surrey (1978), 'Estimating UK Energy Demand for the year 2000—A Sectoral Approach', *Occasional Paper, 5*, Science Policy Research Unit, University of Sussex.

Collingridge, D. (1980), *The Social Control of Technology*, Frances Pinter, London.

Collingridge, D. (1983), 'Hedging and Flexing—2 Ways of Choosing Under Ignorance', *Technological Forecasting and Social Change, 23*, 161–72.

Department of Energy (1978), *Energy Forecasting Methodology*, Energy Paper 29, HMSO, London.

Energy Policy (1978)—A Consultative Document, HMSO, London.

Greenwood, T. (1975), *Making the MIRV—A Study of Defence Decision Making*, Ballinger, New York.

Leach, G., *et al.* (1979), *A Low Energy Strategy for the UK*, Science Reviews, London.

Lewis, C. (1979), 'A Low Energy Option for the UK', *Energy Policy, 7*, 131–48.

Lovins, A. (1979), 'Re-Examining the Nature of the ECE Energy Problem', *Energy Policy, 7*, 179–98.

National Academy of Science (1980), *Energy in Transition 1985–2010, The Final Report of the Committee on Nuclear and Alternative Energy Systems (CONAES)*, Freeman, San Francisco.

Nicholson, R. and A. Farmer (1979), 'The Introduction of Fast Breeder Reactors for Energy Supply', *Uranium and Nuclear Energy*; Proceedings of the 4th International Symposium, Uranium Institute, London, Mining Journal Books, London.

OECD (1979), *Uranium Resources, Production and Demand*, OECD and IAEA, Paris.

Patterson, W. (1977), *The Fissile Society*, Earth Resources Research, London.

Radetzki, M. (1981), *Uranium—A Strategic Source of Energy*, Croom Helm, London.

Royal Commission on Environmental Pollution (1976), *6th Report— Nuclear Power and the Environment*, HMSO, London.

Tammen, R. (1973), *MIRV and the Arms Race—An Interpretation of Defence Strategy*, Praeger, New York.

World Energy Conference (1978), *World Energy Resources 1985–2020: Nuclear Resources*, IPC, London.

York, H. (1975), 'The Origins of MIRV', in D. Carlton and C. Schaerf (eds.), *The Dynamics of the Arms Race*, Croom Helm, London.

York, H. (1976), *The Advisors—Oppenheimer Teller and the Superbomb*, Freeman, San Francisco.

11 FLEXIBILITY IN TACTICAL CHOICE

Chapter 9 introduced the concept of strategic choice. All options which appear to be able to meet the required objective, remembering that it must be stated in general terms, are divided into a number of strategies consisting of options which are variations on a particular technological theme. In the example discussed, the objective was to meet future demand for useful energy and one strategy was the building of additional fossil fuelled generating plant. This strategy consisted of all feasible variations of such plant, built in all realistic programmes. A second strategy was the building of additional thermal nuclear plant, and again this encompassed variations covering all feasible thermal reactors built in all reasonably economic ways. The problem of strategic choice is the problem of selecting one or more strategies as worthy of R & D over fifteen to fifty years or so. It is to ask what kind of technologies are going to be in demand over this time in the future, so that work on their development may commence now. The very nature of this question precludes detailed knowledge about the options which are open; strategic choice is, of necessity, made before all the technologies involved have been developed to a level where they are even partially understood. Some of the technologies may be very familiar, as in the case of the fossil generation of electricity, but this has to be compared with other technologies which are very much less developed and understood only in outline, such as thermal nuclear plants or breeder reactors. Strategic choice is therefore made under very great uncertainty, or ignorance.

The problem of choosing strategies is compounded because attempts to judge between the costs and benefits of options from different strategies involve non-incremental comparisons. Options from the same strategy may be compared more easily because they are similar in a number of ways enabling analytic attention to be directed to the relatively small differences between the options. In comparing across strategies this is not possible, for the options will differ in far more ways than they resemble one other; more or less the entire costs and benefits of one must be put beside a similarly full account of the other's costs and benefits, a task which is generally beyond the available data and often beyond human calculation. It is here that flexibility comes into its own because the flexibility of options drawn from different strategies may be compared. It may be possible to show that all the technological options within a strategy will be inflexible, whatever their details and fine points. Thus it was shown that thermal nuclear power plants have certain features which make them highly inflexible. Moreover, these features are so general that they will apply to all types of thermal reactor and to all programmes for their construction. The features are, it will be remembered, high capital cost and capital intensity, long lead time, large unit size and dependence upon special infrastructure. Breeder reactors, of whatever type, will possess these features to an even more marked degree, so that all options in which energy is provided by breeder reactors will be less flexible than corresponding programmes of thermal plants. In this way strategies may be compared for the flexibility of their options, even at the stage of strategic choice where little may be known of the details of the options.

It is one thing to show that the flexibility of options but not their costs and benefits may be measured across strategies, but it is another to show that flexibility is a desirable characteristic. Showing this, however, only requires reflection on the great uncertainty under which strategic choices must be made. There is, by the very nature of the choice, inadequate information to identify which technologies will be in demand so far in the future. The question is not 'having learned all we can about these technologies, which should be developed further?', but rather 'which of these technologies is it worth learning about?' This immediately places a premium upon technologies which can be learned about quickly and with little cost, but this is to say technologies which are flexible. If a flexible technology is developed, data about its various costs and benefits will be acquired quickly and cheaply, and if it is found to be unsatisfactory, the costs of learning about this mistake will be low as well. If all the technological options within a strategy are inflexible, then this is a reason for not investing in learning about the technology, for such learning will be slow and may be very costly. This, it was argued, was the case for thermal nuclear technology, and even more for breeder reactors. The strategies of building thermal plants and of building breeders may both be stigmatised as *inflexible strategies*. Unless there are special arguments for choosing to develop these technologies, they ought to be rejected in any strategic choice in which they figure.

This is the role of flexibility in strategic choice, but consideration of flexibility also enters what may be termed the *tactical choice* of technology. Once a strategic choice has been made, R & D is directed towards the selected technologies and as more is learned about them, it gradually becomes possible to make more and more detailed comparisons between different varieties of the chosen technology. Choice between the varieties therefore becomes more and more informed and many varieties are eliminated until a few reach maturity and begin to diffuse in the familiar way. This is the process of tactical choice between rival technologies of the same general kind. What is now to be considered is the role of flexibility in this kind of decision. Flexibility will be important for the same reason as before: it ensures that learning is rapid and inexpensive and that the errors which are inevitably made under conditions of such uncertainty do not become unbearably expensive. Here we are talking, of course, of comparing the flexibility of options from the same strategy, unlike strategic choice where the flexibility of options from different strategies is assessed. This means that in making tactical decisions it is possible to trade off flexibility and cost, something which cannot be done in strategic choice where costs and benefits cannot be compared across strategies. This is an important point, for one of the advantages of distinguishing between tactical and strategic choice is that it helps to separate those cases where flexibility can be traded off against cost and those where this cannot be done. A word of warning is necessary, however, before proceeding further. It would be a mistake to think of two kinds of flexibility, strategic and tactical; flexibility is flexibility. It is useful to discuss separately the role of flexibility in making strategic and tactical decisions but in both cases we are concerned with the same concept.

As before, the role of flexibility in tactical choice may be explored by considering the case of nuclear power. In making strategic choices, nuclear technology was condemned as highly inflexible, but some varieties are nevertheless more flexible than others, as has already been shown in comparing thermal and breeder reactors. If, for some special reason, a strategic decision has been made to develop nuclear power, then attention needs to be paid to these differences when tactical choices between different types of nuclear technology are being considered. Flexibility is still a virtue for technologies and although the strategic choice condemns us to suffer a highly inflexible technology, considerations of flexibility in making tactical choices may help to make the best of an unhappy situation. This raises a number of questions. Given that nuclear plant has been selected for development, how can the flexibility of different forms of the technology be compared? How can flexibility be increased? And what is the relationship between flexibility and the normal calculations of costs and benefits, which can be done as the R & D programme continues?

The features which make for inflexibility are high capital costs and capital intensity, long lead time, large unit size and requirements for infrastructure, and these may be varied to alter the degree of flexibility of the technology. Breeder reactors, for example, have been seen to have higher capital costs and capital intensity, longer lead time, larger unit size and greater dependence on infrastructure than thermal nuclear plants, making them even less flexible. Considerations of this sort might lead to the rejection of breeders, even if some form of nuclear plant is to be developed. It may be possible to calculate the cost penalty of abandoning the breeder. If the breeder's benefits under those conditions which are most favourable to it can be calculated, as attempted in the cost–benefit analysis of the USAEC discussed in chapter 1, then this is the greatest loss abandoning the breeder can lead to. Against this is to be balanced the greater flexibility of thermal plant, so that a trade off between cost and flexibility is possible, at least in theory. Within thermal nuclear technology, considerations of flexibility would favour plant designs which were less capital intensive and less dependent upon infrastructure. The CANDU design, for example, uses natural uranium fuel, which requires no reprocessing, the reactor therefore does not depend on enrichment plants or reprocessing plants. This gives it greater flexibility; it can, for example, continue to operate if enrichment technology fails, but its performance may be bettered by reactors using enriched fuel if there are no problems with the enrichment plants which they need. Again, there is a trade off between higher costs in a world which works according to plan and a higher degree of flexibility in order to meet failures of the world to live up to expectations. Flexibility of nuclear plant could also be increased by reducing capital costs and reducing lead times for construction, but these are desirable on simple economic grounds as well, so the reverse may also happen, flexibility increasing as cost decreases.

In the remainder of the chapter attention will be given to the relationship between flexibility and unit size for nuclear plant. This is a general problem since there is clearly conflict between flexibility from small units and economies from large units for any technology which is thought to possess economies of scale. But despite its general importance this topic has received very inadequate treatment to date. The diseconomies of scale are generally much less obvious than the economies to be derived from large units, thus it often happens that plants are built on too large a scale. Accordingly, Abdulkarim and Lucas (1977), Fisher (1978) and Lee (1978) all argue that fossil power stations have been built in units which are far too large. The scale economies of nuclear plant are particularly obvious because of the plants' capital intensity, which makes generating costs very sensitive to capital costs. The unit size of fossil plants has, of course, increased enormously since the early days of the generating industry, but it is important to remember that the pressure for increases in scale here was much less than for nuclear plant. Whatever savings were possible in the capital costs of fossil plants were worth making, but their contribution to reduced generating costs was much less than comparable capital savings for nuclear plant. It was recognised very early on that if nuclear power was to compete with fossil generation, then economies from large plants were essential, and very little attention was given to the diseconomies which might be involved.

The following discussion, which closely follows Collingridge (1980), chapter 7, lists some of the chief diseconomies of scale which can be expected to be found for any technology, and this provides a framework for the later examination of the scale of LWRs in the United States.

(i) *Learning about performance*   The first general diseconomy of scale is that the rate of learning about the performance of a new technology decreases as unit size increases, for the simple reason that fewer units are constructed and operating experience accumulates slowly. We have seen that this is particularly important for nuclear plants because their capital intensity makes generating costs sensitive to capital cost and load factor, two items which are only known once the plant is built and running smoothly. Building in large units means that data on generating costs are only acquired slowly. The same is true of safety, large units implying a slow rate of learning about the technology's safety.

(ii) *Lead time*   The second general source of scale diseconomies is that the lead time of plant tends to increase with its size. This has three consequences. First of all, benefits from the plant are delayed, which is a straightforward economic cost of scale. In addition, a lengthening of lead times requires forecasts about the costs and benefits of the plant to be made over a longer horizon, with a reduction in accuracy of the forecasts and a corresponding increase in the likelihood of investment errors. Thirdly, long lead times delay the acquisition of learning about the technology's performance. Mooz (1978) reports that an increase in size of 1MW in LWRs in the United States corresponds to an increase in lead time of four months.

(iii) *Learning about construction*   The rate of learning about the construction of plant must also decrease with unit size. Mooz (1979) shows that there is significant learning of this sort for United States' LWRs, reporting a reduction in capital costs of 10 per cent and a similar reduction in construction time with each doubling of the number of reactors built by an architect-engineer.

(iv) *Safety*   For hazardous technologies the dangers posed by a large unit can be expected to be greater than those from a small one, if only because the former contains more dangerous material. There has been a tendency for large nuclear units to bear heavier costs in safeguarding dangerous failings (Komanoff (1981a)).

(v) *Bias of future investment* The building of large plants may produce a bias in the following investment round. This is especially true for capital intensive plants and has been discussed before in the case of large, capital intensive plants forming part of the infrastructure for nuclear technology. In the hope of achieving economies of scale, many reprocessing plants, for example, have been built, or are planned, on a very large scale. Their capital intensity means that once built, little is saved by not using them, so in the next round of investment an advantage is given to nuclear reactors having a fuel cycle which can use the reprocessing plant. Large-scale reprocessing plants may exhibit economies of scale, but at the same time they serve to entrench various kinds of reactor design making them, perhaps quite by accident, superior to rival reactors. The French reprocessing plant for the Superphenix, for example, is planned to be able to cope with material from five breeders in all. This gives breeders beyond Superphenix a cost advantage over other reactors which might be chosen or over fossil generation. The freedom of decision makers involved in the next round of investment is limited in this way.

(vi) *The flow of earnings* In general, the failure of large units tends to be more costly than the failure of the equivalent capacity of small units. If the cost of operation is independent of the unit size of the plant, and the probability of failure also independent of unit size, then the expected earnings from the entire plant are likewise independent of the size of its units. On these assumptions, the expected earnings from a large number of small units are the same as those from a few large units. But even here there is a difference because the small units give a better cash flow. In the very long run, there is, it is true, nothing to choose between large and small units, but in the shorter term, the quality of cash flow becomes important and this favours small units (Ball (1975)). In the case of nuclear plant, very large units have lower load factors than smaller ones (Surrey and Thomas (1981)), spreading their capital costs over relatively less output, but even if the failure rate was independent of size, this effect would favour the construction of small reactors.

(vii) *The step effect* There are clearly more total capacities which can be made from a small unit than a large one. If C_s and C_L are the capacities of a small and a large unit and n an integer, then there are more numbers equal to nC_s than to nC_L. This means that there are more options about the system's capacity when small units are used, which tends to minimise costs from the mismatch of capacity and demand. The problems caused by large nuclear plants, when it is realised that they have been ordered in error, have arisen several times in the discussion of nuclear history. For example, the British AGR programme consisted of only five large plants; when it became obvious that they were not needed to meet electricity demand there was little room to alter the programme. It could only be reduced in something like 1,250MW steps. The first AGR, Dungeness-B, was soon recognised to have been contracted out to a totally incompetent consortium, Atomic Power Construction (APC). It is sometimes pretended that this is just one of the problems of innovation of such a radical kind; it is hard to judge the competence of constructors when no one has much experience of the work. But if there is nothing ordinary about such a mistake, there is something quite extraordinary about its cost. Because APC was working on such an enormous unit, errors which might have been forgiveable elsewhere became extremely costly. Smaller units would have reduced the cost of this kind of error.

The step effect also means that there are more sites available for small units, an important consideration for intrusive technology like a nuclear reactor. The lower adaptability of large units is also reflected in the difficulties which sheer size has caused for the export of reactors to less developed economies whose distribution grids may be too small to take even one large reactor. The rule of thumb governing such additions is that no single unit should make up more than 10 per cent of the total capacity of the grid (Smart (1982)).

All of these features show how small units make for flexibility. Flexibility exists when learning about a new technology is rapid and inexpensive, or to put it another way, when whatever errors are made in developing the technology are likely to be discovered quickly and cheaply and if such mistakes impose only modest burdens. If small units are built instead of large, then learning about the technology is speedier, data about performance and safety and skills of construction accumulate more quickly because more units are constructed and because construction takes less time than for large units. In addition to the more rapid accumulation of data and skills, other effects operate to reduce the cost of errors made in planning investment in small units. Failure of small units is less expensive than failure of the same capacity of large units because of the earnings effect. Small units lead to a less severe bias in future investment, giving tomorrow's decision makers more freedom of choice; the step effect means that there is more control over the capacity of a system, and the dangers of hazardous technologies are reduced.

To deepen understanding of these factors, the LWR programme in the United States will be discussed, together with a couple of hypothetical variations. It has been observed earlier in chapter 6 that the large units in which the reactors were built in the expectation of large economies of scale led to a very slow accumulation of data about the technology, particularly about capital costs and load factors, and hence generating costs, but also about the reality of the expected scale economies, the time taken to build plants, plant safety and about how to build such complex pieces of equipment efficiently. The first analysis of any thoroughness of capital costs had to wait until Mooz (1978) because until then there was simply insufficient data. It is therefore interesting to consider what would have happened had reactors been built in smaller units. In the half-size case, let us suppose that all commercial plants which were ordered up to the end of 1967 were of half their actual capacity, twice as many being built, construction of twin plants being parallel. In the one-third-size case, three times as many plants as were actually built are constructed, but of one-third their actual capacity, construction of triplet plants again being parallel. In the actual case fifty-two reactors ordered in these years came into commercial operation between 1968 and 1978, and one, Diablo Canyon 1, which will be ignored for present purposes, came on stream in 1982. Table 11.1 gives the number of reactors which actually came on stream between 1966 and 1978, with the numbers which would have entered operation in the two hypothetical cases.

The results employ Mooz' (1978) finding that the lead time of LWRs increases one month for every 25 MW increase in capacity. Information about the reactors is therefore increased

Table 11.1. Actual and Small-scale LWR Programmes

Year	Number of reactors entering service during the year		
	Actual	$\frac{1}{2}$-size	$\frac{1}{3}$-size
1966	0	0	3
1967	0	4	3
1968	2	4	9
1969	2	6	15
1970	3	10	12
1971	5	14	30
1972	8	20	36
1973	7	24	18
1974	11	6	15
1975	6	14	12
1976	3	2	3
1977	4	0	0
1978	1	0	0
TOTAL	52	104	156

by two mechanisms; building small units increases the number of plants in operation, and small units are built more quickly. Figures 11.1 and 11.2 enable the effects of each mechanism to be separated. Most of the increase in information results simply from building two or three times as many units, the effect of reduced lead times being secondary, though significant.

Far more data becomes available much sooner in the half-size case, and even more in the one-third-size case. The data is not the same, of course; it cannot stretch across the same span of capacities and cannot cover the same time period, but it is clear that building small units would have allowed an understanding of capital costs, load factors, safety, building knowledge, lead time and scale economies to have been achieved far earlier than in the actual case. The thirty-eight reactors operational in the actual case by the end of 1974 can be matched against the thirty-eight in the half-size case by the end of 1971, three years early. In the one-third-size case, forty-two reactors are on stream by the end of 1970, four years earlier than the actual case. In the actual case, capacities vary from between 460 to 1,180 MW, spanning between 230 and 590 MW

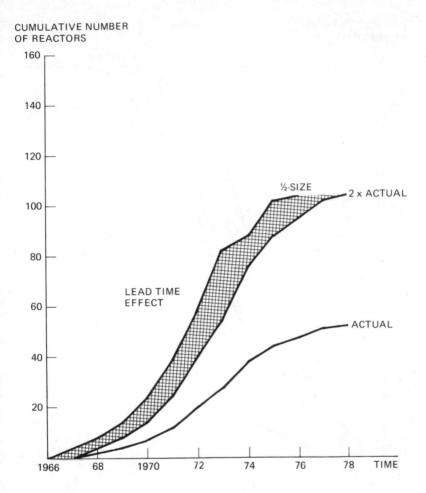

Fig. 11.1. Reactor experience, actual and ½-size cases.

in the half-size case and 150 to 390 MW in the one-third-size case. More data points are therefore needed to reveal scale economies across the smaller range of sizes; but the extrapolation of whatever results are discovered to larger units than have been built would not be a matter of simple mathematics, but would rely on the greater understanding of the components of capital cost in LWRs brought about by the large number which would have been built.

There is obviously a question whether the compressed data

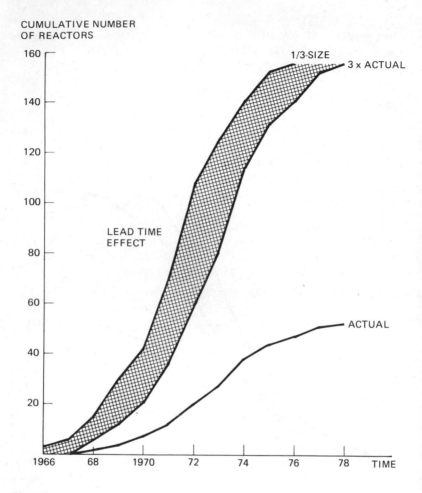

Fig. 11.2. Reactor experience, actual and ⅓-size cases.

points of the hypothetical cases could uncover time-related effects, in particular the very rapid increase in unit capital costs over time reported by Komanoff (1981a) and Mooz (1978, 1979) and discussed in chapter 6. It might be thought that data over many years are required for this, whatever the scale of construction, but this would be wrong because there are no pure time effects on LWR capital costs, load factors, safety or whatever. Time alone alters none of these; if capital costs tend to fall over time, then this is because, for example, architect-engineers acquire building experience over time. If load factors increase during the early years of a nuclear plant, as analysed by Surrey and Thomas (1981), then this is not brought about by the passage of time, but by the labours over time of the plant's operators who seek to improve its performance. The increase in LWR capital costs must be explained in a similar way, by for example the suggestion of Komanoff (1981b). He argues that increasing operating experience leads to the discovery of ever more routes for failure, which must be safeguarded against in future reactors and those in sufficiently early stages of construction, with a consequent rise in their capital costs. But once a mechanism is sought, the problem about compressed data does not look so serious. If Komanoff's suggestion is true for the actual and the half-size cases, then operating experience in the latter accumulates much more rapidly, leading to the earlier discovery of failure modes, the safeguarding of which will force up capital costs even more rapidly than in the actual case. Thus, the compression of data on capital costs in time will be compensated for by a more marked increase in capital costs over time; the other hypothetical case works similarly.

But what of the cost of building small units? The major component of cost here is, of course, the loss of economies of scale. At the beginning of the construction programme there was no operating experience on which to base estimates of economies of scale, but they were forecast to be large and so large units were built. This is the old mistake, discussed in chapter 1, of basing decisions on the best guesses that can be made from the data available at the time. What was forgotten, and what has a cardinal role here, is flexibility. In a state of ignorance about what economies of scale exist it might be thought a matter of throwing a coin to settle for large or small units, but thinking about flexibility destroys this symmetry. If economies of scale are smaller than forecast, as was the case here, building large units in the hope of capturing scale economies means that knowledge of capital costs accumulates very slowly. It therefore takes several years before enough data is available to reveal that scale economies are, in fact, smaller than originally thought. By this time, however, much irrecoverable investment has been made in plants which are too large, so the cost of the erroneous forecast is high. If, on the other hand, scale economies are much larger than forecast, building small units in the light of the forecast means that data about capital costs accumulates more rapidly than before, enabling the error in the forecast to be discovered much earlier. Once there is data showing what the real economies of scale are, larger plants may be ordered from then onwards, so that the cost of error, the foregone scale economies of the early plants, is limited. In the absence of data on economies of scale it therefore makes thoroughly good sense to begin by building small units.

Since the early days of LWR construction data on scale economies has accumulated, and Mooz (1978) reports that an increase in size of 1MW produces the modest decrease in unit capital costs of $ (1978) 0.25/kW, a figure in pretty good agreement with that of Komanoff (1981a). In his later study however, Mooz (1979) was unable to detect any scale economies, but even his earlier result was far lower than the forecasts made in the early days of LWR construction in the United States. Taking Mooz' earlier figure, the extra capital costs of the half-size case are about $ (1978) 4.8 billion, and the one-third-size case has capital costs of $ (1978) 6.4 billion over the actual case. The capital cost of those plants ordered up to the end of 1967 totals about $ (1978) 30 billion, making allowances for the early turnkey contracts which understate the real costs; the capital costs from lost scale economies amount to 16 per cent for the half-size case and 21 per cent for the one-third-size case.

The calculations do not end here, however, for the other advantages of small units must be included in the reckoning. There are two features which offer some compensation for loss of scale economies and which may be quantified easily; reduced lead times give benefits earlier and building more reactors gives greater experience of construction and so decreases capital costs more rapidly. The half-size reactors have an average reduction in lead time of about sixteen months, and the one-third-size twenty-one months. There is no point in attempting to be over precise; accordingly the benefits from this effect will be taken as 10 per cent of capital costs for the half-size case and 15 per cent for the one-third-size case. Mooz (1979) reports that capital costs fall by 10 per cent with each doubling of construction by one architect-engineer. In the half-size case twice as many plants may be assumed to be built by each architect-engineer, reducing total capital costs by 5 per cent. In the one-third-size case, this reduction will be about 8 per cent. Following the analysis of Surrey and Thomas (1981), it is best to assume that load factors for the small units in both hypothetical cases are the same as for the large units of the actual case. They found that plants of under 400MW had lower load factors than larger ones, but it appears that this is because all such plants were early ones, so that the difference is due to learning and is not an effect of scale itself. The two compensations for loss of scale economy reduce the extra capital cost of the half-size case to $ (1978) 4.1 billion and of the one-third-size case to $ (1978) 4.9 billion. It must also be remembered that no allowance has been made for the other benefits of smaller units: a better flow of earnings, improved safety, less bias of future investment and more control over the capacity of a particular system. These are, of course, harder to quantify, but it must be remembered that the final extra costs of the two hypothetical cases is to this extent an over-statement.

The extra costs of acquiring more and earlier information about the performance of nuclear plants may be compared to the investment in question. During the period from 1968 to 1974 average ordering was nineteen plants per year, with an average size of about 1,000 MW. Thus capital investment during this period was running at something like $ (1978) 20 billion per year, a figure necessary if the new technology was to become firmly established. Thus an expenditure of $4 to 5 billion would have been needed to obtain information about a technology which was being invested in at $20 billion per year, three to four years earlier than permitted by the large-scale construction of the actual case. This would seem a trifling sum to pay, and would remain so even if the loss of scale economies were several times as great.

The calculations presented here are very simplified, but they nevertheless make their point. Whatever the complications which have been avoided in these simple sums, it is clear that LWRs in the United States were built on far too large a scale and that the heavy cost of that country's nuclear adventure would have been much less had those designing and investing in it shown a greater sensitivity to the problems of scale. To justify this conclusion it is necessary to avoid charges of using hindsight. When tactical decisions had to be made about the scale of commercial plants in the middle- to late-1960s before the LWR was commercialised, there was data only on early, very small reactors, which could not be extrapolated to larger units without raising all kinds of questions. Any purchaser of a larger commercial reactor was therefore making an investment under very considerable uncertainty, despite the optimistic promises of the suppliers and the USAEC. In such cases it is advisable to consider what would happen should the investment prove mistaken, if, for example, it should turn out that a coal-fired plant would have produced electricity more cheaply. It would need no specialist knowlege of nuclear technology to realise that the reactor's high capital costs and capital intensity mean that little could be done to alleviate a mistaken investment. In other words, any mistake in buying the new technology is likely to prove expensive. Thus caution is easily deduced from some very general features of nuclear technology.

This alone is sufficient reason for favouring small reactors, so that any utility's risk may be kept low. If this were not enough, then there is some urgency in learning about the new technology in order to reduce the risks of error by its pur-chasers. It is therefore necessary to obtain operating experience of commercial plant quickly, and it requires no qualifications in atomic physics or nuclear technology to realise that building small reactors will provide experience more quickly than building large ones; to call this elementary arithmetic would be to exaggerate the issue. There are therefore two sound and completely non-technical reasons for favouring small units: they limit the risk undertaken by purchasers, and they lead to a more rapid reduction in this risk. Against this must be reckoned the loss of scale economies, if indeed these are going to be found to be real. Not knowing what the real economies of scale are it makes sense, as I argued before, to invest in small units. If small units are built, data on performance accumulates more rapidly and the real economies of scale will become apparent sooner. If they are appreciable then there is at least data to justify building larger reactors and the cost of error is limited to the loss of scale economies for the early reactors. If, on the other hand, large units are built, data becomes available in a dribble and many years may go by before it is discovered that the units are too large, by which time a large investment has been made in oversize plants. It should, moreover, have been clear that any loss of scale economies in these early reactors would be small compared to the investment needed for the large-scale substitution of nuclear for fossil plant. If nuclear plant was going to be a success, then orders would have to run at between 20 to 30 GW per year, representing a very large investment, against which any losses from building undersized plants in the early days of the technology is sure to be trifling. If, on the other hand, the technology is going to be uncompetitive with fossil generation, then it makes sense to limit investment to small reactors anyway. If the new technology is going to take off or fail, it therefore makes sense to favour small units. Nothing in these arguments exploits the privilege of hindsight; it may therefore be concluded that investors in LWRs during this period could have, and should have, recognised the benefits to be obtained from buying small reactors. Without using knowledge and insights which were not available to them, it can be said that they invested in LWRs which were too large.

It is now time to return to the general issue of the role of flexibility in the tactical choice of technologies. The example of United States' LWRs concentrates on one feature which contributes to the flexibility of options being considered in tactical choice, namely the scale of the technology's units. The general lessons of the example are that considerations of flexibility are important in tactical choice; that flexibility is enhanced by small units; that in tactical decision making flexibility may be traded against more familiar economic costs; that decision makers can exhibit extraordinary insensitivity to questions of scale and that the same concept of flexibility as applied to strategic choice may be applied to tactical decisions and may be measured in exactly the same way. The last point underscores the warning given earlier about thinking that there are two varieties of flexibility, strategic and tactical. There is not; there is one concept which may be used in choices which are strategic and those which are tactical. Other ways of enhancing flexibility have been discussed in less detail and include reducing capital intensity and dependence upon infrastructure.

In the discussion so far no use has been found for numerical measures of flexibility, but there are many suggestions in the literature as to how flexibility, or whatever term may be chosen to cover the concept, may be quantified in this way. In strategic choice between technologies there seems to me to be little use for such measures because they all require more information than it is generally realistic to ask for. In strategic decision making, therefore, the non-numerical approach of chapter 9 seems to be required. But the same is not the case for all tactical decisions. Much more information is available for tactical than strategic decisions, so it may happen in a particular case that enough data exists to use some numerical measure of flexibility. The literature on this is surprisingly large, if not to say enormous, but its diversity has hidden it from any but the most determined of researchers. I am therefore most grateful to the persistence of my student Dr. Stuart Evans (1982a, 1982b) in drawing together such a diffuse body of work. The most familiar measures of this sort are Rosenhead's robustness (Rosenhead and Gupta (1978), Rosenhead *et al.* (1972)), Pye's (1978) entropy measure and measures based on the size of the choice set (Marshak and Nelson (1962), Merkhofer (1977)) and the economic approach of Stigler (1939). But in addition, there are many other measures developed to suit particular problems and tailored to the kind and quality of information available. Thus, to give just a few selections from a very wide body of literature, Kerchner (1966) examines the flexibility of dairy plants and Collins (1956) has studied the flexibility of grain production; Fumas and Whinston (1979) have applied the concept of flexibility to contracting theory; Friedman and Reklaitis (1975) and Van der Vet (1977) have examined flexibility in linear programming. Flexibility has also been studied for the choice of programming systems (Grinyer and Wooller (1978) and Mitchell (1970)), and in planning tele-communications (Montenegro (1977)). The problem of flex-ibility of water resources has also received attention (Fiering (1982a, 1982b, 1982c) and Hashimoto *et al.* (1982a, 1982b)). The list could be extended, but I hope that enough has been said to show that there is a very valuable body of literature under development which should receive more attention.

References

Adbulkarim, A. and N. Lucas (1977), 'Economies of Scale in Electricity Generation in the UK', *Energy Research*, **1**, 223-31.

Ball, D. (1975), 'Flixborough—Too Many Eggs in One Basket?', *Process Engineering*, August, 55-9.

Collingridge, D. (1980), *The Social Control of Technology*; Frances Pinter, London.

Collins, N. (1956), *Gains from Flexible as Compared with Inflexible Use of Resources*, Ph.D. thesis, Harvard University, Cambridge, Mass.

Evans, S. (1982a), *Flexibility in Policy Formation*, Ph.D. thesis, University of Aston in Birmingham, UK.

Evans, S. (1982b), *Strategic Flexibility in Business*, SRI International, Business Intelligence Program, Report No. 678.

Fiering, M. (1982a), 'A Screening Model to Quantify Resilience', *Water Resources Research*, **18**, 27-32.

Fiering, M. (1982b), 'Alternative Indices of Resilience', *Water Resources Research*, **18**, 33-9.

Fiering, M. (1982c), 'Estimates of Resilience Indices by Simulation', *Water Resources Research*, **18**, 41-50.

Fisher, J. (1978), 'Economies of Scale in Electric Power Generation', International Institute for Applied Systems Analysis, Seminar 10 October, IASSA, Laxenburg, Austria.

Friedman, Y. and G. Reklaitis (1975), 'Flexible Solutions to Linear Programs Under Uncertainty', *AICHE Journal*, **21**, 77-89.

Fumas, V. and A. Whinston (1979), 'Flexible Contracting Theory and Case Examples', *European Journal of Operational Research*, **3**, 368-78.

Grinyer, P. and G. Wooller (1978), *Corporate Models*, Institute of Cost and Management Accountants, London.

Hashimoto, T., J. Stedinger and D. Loucks (1982a), 'Reliability, Resiliency and Vulnerability Criteria for Water Resources Systems Performance Evaluation', *Water Resources Research*, **18**, 15-21.

Hashimoto, T., D. Loucks and J. Stedinger (1982b), 'Robustness of Water Resources Systems', *Water Resources Research*, **18**, 21-6.

Kerchner, O. (1966), Economic Comparison of Flexible and Specialized Plant, Ph.D. thesis, University of Minnesota, USA.

Komanoff, C. (1981a), *Power Plant Cost Escalation: Nuclear and Coal Capital Costs, Regulation and Economics*, Komanoff Energy Associates, KEA-12, New York.

Komanoff, C. (1981b), 'Sources of Nuclear Regulatory Requirements', *Nuclear Safety*, **22**, 435-48.

Lee, T. (1978), 'Optimization of Size in Power Generation', International Institute for Applied Systems Analysis, Seminar, 5 December, IIASA, Laxenburg, Austria.

Marschak, T. and R. Nelson (1962) 'Flexibility, Uncertainty and Economic Theory', *Metroeconomica*, **14**, 42–58.

Merkhofer, M. (1977), 'The Value of Information Given Decision Flexibility', *Management Science*, **23**, 716–27.

Mitchell, J. (1970), *The Design and Construction of Flexible and Interactive Programming Systems*, Ph.D. thesis, Carnegie-Mellon University, USA.

Montenegro, J. (1977), *Planning Communications Systems*, Ph.D. thesis, The Wharton School, University of Pennsylvania, USA.

Mooz, W. (1978), *Cost Analysis of Light Water Reactor Power Plants*, Rand Corporation, R–2304–DOE, Santa Monica.

Mooz, W. (1979), *A Second Cost Analysis of Light Water Reactor Power Plants*, Rand Corporation, R–2504–RC, Santa Monica.

Pye, R. (1978), 'A Formal Decision Theoretic Approach to Flexibility and Robustness', *Operational Research Quarterly*, **29**, 215–29.

Rosenhead, J. and S. Gupta (1968), 'Robustness in Sequential Investment Decisions', *Management Science*, **15**, 18–29.

Rosenhead, J., M. Elton and S. Gupta (1972), 'Robustness and Optimality as a Criterion for Strategic Decisions', *Operational Research Quarterly*, **23**, 413–28.

Smart, I. (1982), 'The Consideration of Nuclear Power', in J. Katz and O. Marwah (eds.), *Nuclear Power in Developing Countries*, Lexington Books, Lexington, Mass.

Stigler, G. (1939), 'Production and Distribution in the Short Run', *Journal of Political Economy*, **47**, 305–27.

Surrey, J. and S. Thomas (1981), 'Worldwide Nuclear Plant Performance', Appendix 5 of House of Commons Select Committee on Energy, *Report on the Government's Statement on the New Nuclear Power Programme*, Vol. **3**, HC 114-iii, 781–804.

Van der Vet, R. (1977), 'Flexible Solutions to Systems of Linear Inequalities', *European Journal of Operational Research*, **1**, 247–54.

12   THE BOUNDARY PROBLEM

1. The boundary problem

Synoptic rationality tries to show how decisions may be justified, but its prescriptions remain ideal for all but the most humble decisions because following them requires obtaining vast quantities of information. All aspects of the decision problem which might conceivably be relevant need to be considered, for only a comprehensive view can show which is the best choice. Trying to apply the rules of synoptic rationality pushes the boundary of the problem further and further away, so that it soon becomes quite unrealistic. This is, of course, recognised by the proponents of this view of decision making, who pretend to cope with the problem by presenting their rules of rationality as ideals which, if they cannot be followed in practice, can at least be aimed at. We have seen, however, that this is just a play of words because the synoptic ideal is completely silent about how decisions should be made where the ideal cannot be met, making such decisions arbitrary.

A good example was provided by the cost-benefit analysis of the breeder reactor considered in chapter 1, especially by its treatment of uranium prices which are, of course, crucial to the development of breeder technology. The study attempted to limit its task by considering only the supply and demand for uranium in the United States. This was difficult enough because supply depends upon the distribution of ores throughout the country, about which little is known, and because demand is determined by the rate at which thermal reactors are built in the country, which is very difficult to forecast. But trying to bound the problem in this way does not work. Uranium is traded internationally, so that calculation of the benefits of using it efficiently in breeder reactors depends on the world prices for the metal, not those in America. Estimating world supply and demand is therefore necessary, which requires the vastly more difficult tasks of discovering the distribution of ores across the world and of forecasting the growth of world thermal nuclear capacity. In this way the boundary of the problem becomes extended in an attempt to meet the strictures of synoptic rationality to an extent where it is impossible to handle. If practical constraints on money, brains and time mean that the boundary has to be drawn more narrowly, how can accusations of arbitrariness be avoided? In the present case, for example, uranium supply and demand in the United States was thought to be a realistic, manageable boundary which can be reached without excessive expenditure and delay, but without some reasons beyond practicality for selecting this boundary, it can only appear arbitrary. It is too much to hope that a rationally defensible boundary will just happen to co-incide with what is practical in a difficult and uncooperative world. Synoptic rationality presents decision makers with a dilemma: they may set about collecting a quite impractical quantity of information, or they may choose in a way which is arbitrary.

The problem does not arise for incrementalist views of decision making, which is hardly surprising since they were developed as a reaction to the failings of synoptic rationality. In Lindblom's partisan mutual adjustment, for example, each partisan is concerned with a narrow aspect of a policy problem and need make no attempt to grasp more than this. Even with this narrow interest, the problems of analysis are simplified by the partisans' concern with the immediate consequences of only a handful of options, so chosen that incremental comparison is possible. In this way, each partisan has to tackle a decision problem which is bounded in a manageable way. I have argued that such views of decision making are perfectly in order provided that the options being considered are flexible. Flexibility enables the partisans to adjust to one another's interests in a wide variety of ways so that they can exercise control over the decision. If, for example, one partisan discovers some unexpected ill-effect of an earlier decision, there will be many ways in which his fellow partisans can adjust to ameliorate the problem, making its control straightforward and rapid. If, however, an extremely inflexible decision is made, this will not be possible. Whatever bad consequences are discovered, there will be little room for control, and their amelioration is likely to be very partial, time consuming and the source of great friction between partisans. The discussion of these problems has been based around nuclear power as an example of a highly inflexible investment, but the lessons learned are all general. The most important of these is that incremental decision making can be efficient and can work in the way described by Lindblom and others, only if inflexible investments are avoided.

This offers a back door route for the boundary problem to reappear. If incremental decision making calls for attention to be paid to flexibility, and if the measurement of flexibility requires information about many aspects of the decision problem, then the boundary of the problem may expand to limits which are impractical, thus incrementalism is as much troubled by the boundary problem as synoptic rationality. It is the purpose of this chapter to show that flexibility can be incorporated into incrementalism without the reappearance of the boundary problem.

At first sight the problem may not appear very pressing because the way in which it is proposed to measure flexibility restricts the decision problem very narrowly by looking at the system itself rather than its interaction with the environment. Any decision may be thought of as a choice between a number of systems, the payoff over time from a system being determined by its interaction with the environment. For all cases outside textbooks, it is impossible to predict the payoff from a system because it is usually very difficult to forecast the behaviour of the environment, and because the way in which the system interacts with the changing environment is generally very poorly understood. There is therefore plenty of room for surprises and unexpected shocks. It is impossible to pick the system which is going to have the highest payoff over time, but this does not mean that choice must be arbitrary. A system will have a finite number of decision variables which may be changed over a certain range at a particular speed and a particular cost, in response to information about the environment which takes a certain time to be recognised. The payoff from the system may be altered by changing decision variables, although there is no simple correlation between these and the payoff, since this is mediated by the system's interaction with the environment. If a system has many decision variables which may be changed over a wide range quickly and cheaply, and if information about the system's payoff is received quickly, then the system has the virtue of flexibility. Such a system will reveal error, in the shape of an inadequate payoff, quickly. Because of the number of decision variables and their ranges, there will be many remedies for the error once it has been discovered, and since these are quick and cheap to operate, the payoff is likely to be improved rapidly. The system is easy to control. Putting it formally, the system has a low monitor's response time, a low corrective response time, a low control cost and a low error cost. The monitor's response time is the time taken to detect bad performance; corrective response time, the time taken to remedy a bad performance; control cost is the cost of changing decision variables and error cost is the loss of payoff, which is low because of its speedy remedy. A flexible system is therefore likely to perform better in the long run than a system of lower flexibility. To put it another way, learning about the performance and control of a flexible system will be quicker and cheaper than learning about a less flexible system; over the long run its payoff is likely to be greater.

The idea here is that trying to forecast the system's behaviour over time is impossible. All sorts of quite unexpected things may happen which would call for the system to be adjusted. There is thus no way of telling what problems a system may run into during its lifetime. But while the potential problems are all too numerous, the responses which the system can make are very often few. In driving a car, there are countless events which might call for it to change speed or direction, but the responses to all these events are brought about by a gear lever, a brake pedal, a clutch and an accelerator. In considering which system to choose there is no hope in trying to forecast *why* it will need to be changed, but *how* it can be changed is often a very simple matter. If one car has a smaller turning circle than another, it can immediately be pronounced more steerable, although it is impossible to predict just when its smaller circle will be useful. Flexibility is assessed by considering not how the system will interact with its environment, but by looking simply at how the system may be altered. The environment may be forgotten as far as the measurement of flexibility is concerned, so that the exact distinction between system and environment should not affect the measurement. This is what is hoped for: thus measuring flexibility does not involve an impractical widening of the decision problem's boundary.

A deeper inspection, however, raises doubts about this happy conclusion. Consider, for example, a power station which may be fuelled by coal or oil. If the station itself is considered, it appears to be very flexible because it can quickly respond to changes in the relative prices of the two fuels, burning whichever is the cheaper. But if the station is considered together with the supply of its fuels, then it appears much less flexible. If there is a sudden rise in oil prices, making coal the cheaper fuel, there may be problems in supplying coal because of the very slow rate at which coal output can be increased. The speed and expense of the switch from oil to coal therefore look very different if the station is considered by itself or together with its fuel supplies. It would seem, therefore, that the flexibility of a system may change as the boundary of the system is expanded, so that flexibility depends on the way the system is distinguished from its environment. But if flexibility is relative to the boundary, how can the boundary be drawn in a way which is not arbitrary? It looks as if we have regenerated the old dilemma between drawing a practical, but arbitrary, boundary around the system and expanding the boundary until it encompasses everything. Adding flexibility to incrementalism seems to run into the same boundary problem which counts so heavily against synoptic rationality.

In reply to this, I shall argue that expanding a system's boundary can never lead to an increase in flexibility. In the case of the dual-fired power station, this means that expanding the system to include its fuel supply cannot increase the flexibility of the plant; it can only decrease it or leave it unaltered. The rate at which the dual-fired station may be converted from oil to coal is obviously limited by features of the plant itself, as is the cost of conversion. If it takes a month to adjust the boiler and if this costs £100,000, then the conversion to coal cannot possibly take less than a month, nor cost less than £100,000. Problems in supplying coal may greatly lengthen the time needed for conversion and other factors may greatly increase its cost but they cannot speed up or cheapen the change. In this case, therefore, flexibility cannot be increased in expanding the system under consideration from the dual-fired station itself to the station together with its fuel supplies. If this can be shown to be general, then the boundary problem may not be so daunting. Let us see if it is.

How then, at a general level, can changing the boundary of a system alter the measurement of its flexibility? This can happen by changes in time or cost, namely the monitor's response time and the corrective response time and control costs and error cost. In the case of time, no expansion of the system's boundary can reduce either of the response times. If the nature of the system means that it takes time m to detect some error signal and time c for its decision variables to be adjusted to correct the error, then the system itself constrains the speed of error detection to be greater than or equal to m, and the time for response to error detection to be greater than or equal to c. This remains the case when the boundary of the system is expanded; nothing which this takes from the original environment and places in the larger system can serve to reduce either time. In the case of the dual-fired boiler, for example, suppose it takes three months to detect any significant trend in oil prices, the monitor's response time, and it takes a month to adjust the boiler from oil to coal burning, the corrective response time. Expanding the system to include the supply of fuel as well cannot do anything to lessen these times. If it takes six months to detect trends in coal prices and if coal output could be increased to supply fuel in nine months, then identifying the need for conversion might be delayed and it might take much longer to acquire coal in response to the detected need. But if trends in coal prices can be detected in one month, and coal supplied within a week, the speed of change is limited to its original value by the need to detect significant trends in oil prices and to convert the boiler. Or think of a car and all the movements which are dynamically possible to keep the vehicle on the road. If the system is now expanded to include the driver, who may have poor eyesight and slow reactions, the range of movements which are dynamically possible is obviously diminished. A deceleration which is possible for the car itself may be rendered impossible by the weakness of the driver's right foot, and panic may prevent him from correcting a skid which is open to correction when just the forces acting on the vehicle are considered. The small system constrains the performance of the larger; nothing in the larger can overcome these constraints and improve the performance of the small.

The same is true for costs. If the control cost and error cost for a system have been properly identified, then expanding the system's boundary cannot reduce these costs; it may only leave them unchanged or increase them. As the boundary of the system is expanded new varieties of cost may be uncovered, but the original ones remain unchanged, so that total costs cannot decrease. If extra costs are discovered, then the control cost or the error cost increases, decreasing the flexibility of the system. Since the costs originally identified persist, the system's flexibility cannot in this way be increased. Earlier discussions of nuclear power, for example, considered just the technology as nuts and bolts, costing so much and taking so long to build. Considered in this way, nuclear technology was seen to be highly inflexible. Later on, other costs were considered, such as disruption to the environment, hazards to health and life and the proliferation of nuclear weapons. There is some doubt about the reality of these costs, but if they are neglected in decisions to invest in nuclear power and if they later prove to be real then they must be added to the cost of error as originally calculated. Error costs, in other words, may be increased by considering new varieties of cost, but they cannot be decreased. The flexibility of nuclear power cannot therefore be increased by expanding the system boundary to include other kinds of cost; flexibility remains the same or it decreases.

So far, then, it has been argued that the problem of flexibility varying with the boundary placed round a system is ameliorated by the fact that expanding boundaries cannot increase flexibility. Wherever the boundary is drawn will yield a maximum estimate for the system's flexibilty, which will either remain unaltered or be reduced as the boundary of the system expands. As I have tried to show throughout this book, there is a need to identify inflexible options in policy making, and to consider what problems they might generate and how their flexibility might be improved. If the flexibility of a particular system is low, then it must be remembered that it can only get lower as the system boundary is expanded. There is always a danger, how-ever, of finding a system which appears quite flexible, but which would appear very much less so were its boundary to be expanded in a particular way. The warning is that judge-ments of high flexibility must be treated as provisional, and where appropriate must be tested by expanding the boundary of the system to see if flexibility is thereby reduced. Just how serious a problem this is in real decision making must await practical applications.

2. Nuclear power and fuel diversity

The somewhat theoretical discussion of the chapter so far may be made a little more real by considering an argument for nuclear power which has become prominent at the public inquiry into the construction of a 1,200MW PWR at Sizewell in England. The case of the CEGB, who is applying for permission to build the reactor, is a double one. There is considerable overcapacity on the CEGB's system, which covers England and Wales, so that the new reactor is not needed to meet electricity demand. Nevertheless, the CEGB argues, it will be worthwhile because it will displace old coal plant whose very high generating costs are expected to worsen as coal prices rise in the future. The Board's other argument is that even if the reactor were not an economic investment, it would be worth building as a means of diversifying the fuel supply of the Board's generating plant. Fuel diversity is supposed to have two benefits. First of all it protects the Board's customers against sudden shocks from unexpected changes in fuel prices, as occurred in 1973 and 1979. CEGB (1982b), 23 states: 'Diversification of the types and sources of fuel used for electricity generation helps to reduce the consequences of a shortage in any one fuel source, or a price rise in any one fuel source'.

The second benefit is that fuel diversity reduces operating costs, which is itself a double effect. It helps to keep prices down by preventing a monopoly fuel supplier and it enables the pattern of fuel consumption to be adjusted to changes in the relative prices of the available fuels. As CEGB (1982a), 30 tells us:

Over the lifetime of a station there can be considerable variations in the type of plant which provides the lowest operating costs, because of unexpected fuel price changes—as has happened with oil. Differential fuel price increases will occur partly at least in response to their differing availabilities in the future. The CEGB accordingly seeks to have a reasonable balance on its system between generating plant using different fuels. In that way, the consequences of undue dependence on a single fuel can be avoided or lessened and the consumer partly shielded from abrupt price changes.

At the moment the CEGB is heavily dependent on coal which provided over 80 per cent of its electricity between 1981 and 1982, accounting for 58 per cent of revenue. If the ageing Magnox nuclear plants are not replaced by new nuclear plants, then the dependence on coal will improve only slightly. Nuclear plants accounted for 11 per cent of generation during 1981 to 1982, and this will creep up to 14 per cent when the four AGRs are on line, allowing for retirement of old Magnox stations. Diversity therefore seems to be needed, but the CEGB's studies show that oil prices are likely to continue at their present high level, and that electricity from renewable energy sources cannot be relied upon to provide anything more than a fraction of total generation over the next fifty years. Thus, 'the clear implication is that new nuclear plant represents the only viable large scale alternative to coal based generation at the present time' (CEGB (1982a), 30).

Various scenarios for economic growth were used in the CEGB's analysis, with primary energy needed for electricity generation in 2030, the study's planning horizon, varying from 78 to 233 mtce. No more than 20 per cent of this can realistically be provided by renewable energy, even by 2030, and if oil remains expensive over this period the Board will have to meet 80 to 100 per cent of its fuel needs from coal. With other demands for coal, there are obviously many future conditions under which the United Kingdom would revert to being a net importer of energy. This opens the way for sudden price shocks such as those of 1973 and 1979. These are to be expected because world fossil fuel markets are likely to be highly imperfect, with cartels, long lead times for new development and delays on environmental grounds. The dismal story is told in CEGB (1982b), 46:

Uncertainty arises in particular from the commercial combination of a small number of major exporting countries coupled with long lead times for the development of new coal and many new oil sources, with the possibility of delays caused by environmental considerations. The long lead times in particular can easily lead to a mis-timing of necessary investment and price instability in addition to constraints on marginal availability. Although the greater uncertainty attaches to imported fuels, indigenous supplies are not without their own uncertainties. The uncertainty in the output from renewable energy resources is evident from the foregoing discussion and environmental considerations can delay new UK coalfield developments as well as foreign ones.

There is therefore great danger in any utility continuing to rely on fossil fuel as heavily as the CEGB relies upon coal. If nuclear power were developed as far as possible in the CEGB's system, then it could provide between 71 and 207 mtce of primary energy by 2030, depending on which scenario is used. This would reduce substantially the Board's dependence on coal, making it easier for the country as a whole to remain self-sufficient in energy. Uranium would, of course, have to be imported, but it is not likely to be open to the kind of unexpected shocks which are likely to afflict markets for fossil fuels. Nuclear generating costs are not particularly sensitive to the price of uranium, limiting the risks from increases in uranium prices in any case.

This is the CEGB's case for the Sizewell PWR on grounds of fuel diversity. It is of interest to us because it would appear that nuclear power is being appealed to as a provider of flexibility. Nuclear power, it is argued, promotes the flexibility of the CEGB's generating system by relieving monopoly pressure, making price shocks less likely to occur, and by making the system less sensitive to whatever shocks do occur by enabling fuels to be substituted and by spreading risks. The performance of the system with nuclear power seems to be less sensitive to errors in forecasts about fossil fuel prices than the system with no nuclear power, so that the cost of erroneous forecasts is likely to be lessened. This means that the system with nuclear power is the more flexible. Here, then, is the problem. In considering nuclear power up to this point, attention has been paid to the nuts and bolts of the technology, such as how capital intensive nuclear plant is, how long it takes to build, what size its units are and what infrastructure it needs. With this view, it has been argued that nuclear power is not desirable on grounds of its great inflexibility. But now we have no less than the CEGB telling us that when the boundary of the system is expanded to encompass fuel supplies as well as the hardware for burning them, then nuclear power actually enhances flexibility. This is, of course, quite contrary to the earlier discussion of this chapter where it was argued that no expansion of the system's boundary can lead to an increase in flexibility. Either the CEGB is wrong, or else these earlier arguments are in error and flexibility must be admitted to be dependent on where the system's boundary is drawn and therefore much diminished as a useful tool in decision making. It is therefore necessary to consider the CEGB's case in a little more depth. I shall argue, contrary to the Board, that nuclear technology is inflexible, and that it is therefore a very poor way of getting whatever is required, be it cheap electricity or fuel diversity.

Is nuclear power really an efficient way of diversifying fuel supplies for the CEGB's electricity generation? The central problem is that of shifting from one fuel to another. For oil- and coal-fired plants it is easy for the system to accommodate changes in relative prices; it merely needs a recalculation of merit order and new quantities of fuels to be ordered. Thus, as oil prices have increased relative to coal, the CEGB has shifted away from oil until much modern oil-fired plant is at the bottom of the merit order and the oil burn is insignificant. This is a reflection of the flexibility of fossil plants conferred by their relatively low capital intensity. It is quite another story for nuclear power, however, as we have seen many times before. The high capital intensity of these plants means that little is saved by not running them, making it very costly to operate them at anything less than their full availability. Nuclear plant is always top of the merit order ahead of fossil plants because merit is a reflection not of generating costs but of fuel costs. In a generating system of the sort envisaged by the CEGB, with most fuel being either nuclear or coal, the former would operate on base load, with coal providing the remainder of the base load and non-base load. If the price of coal rises it is therefore impossible to shift away from coal to nuclear power immediately because existing nuclear stations will be working to their full availability. Similarly, if coal prices fall to make its generating costs lower than those of nuclear, there cannot be an immediate shift away from nuclear power to coal because this would leave capital intensive nuclear plants working at less than full availability. Any shift between the fuels has to await the construction of new plant; it cannot be done in the straightforward way coal has displaced oil over the past decade. Building new plant takes about ten years, requiring at least this time for the system to respond to changes in coal prices.

What effect does this have on the CEGB's arguments that nuclear power confers fuel diversity? Consider first the argument that nuclear power protects electricity consumers against sudden and unexpected price increases or shortages of fuel. There are three ways in which this benefit is conferred. Firstly, the diversity of fuel means that the CEGB is protected from the worst kind of shock, that which comes from sudden price increases or shortages of the single fuel used by a utility. Nuclear power certainly enables these shocks to be avoided, although there might be better ways of achieving the same result. The second protection which nuclear power is supposed to confer is that there can be a shift between fuels as their relative prices change, limiting generating costs. This is true for coal or oil where the plant involved is of similar and fairly low capital intensity, but as we have seen, it is not true of nuclear power and coal where changing the fuel mix cannot be done in less than ten years and very often takes much longer, if, for example, there is little increase in demand and few retirements of old plant. Thirdly, nuclear power is supposed to make sudden price rises and shortages of coal less likely because it prevents monopoly pressure from the National Coal Board (NCB). But the inability to shift rapidly between coal and nuclear power greatly weakens this benefit. Monopoly pressure can be effectively resisted only if the fuel supplier knows that any increase in price will entail a smaller purchase of his fuel in favour of its rivals. In the case of nuclear power and coal, this cannot happen in the short term because if coal prices increase, there cannot be a shift away from coal to nuclear for at least ten years, and probably much longer. Monopoly pressure from the NCB cannot be resisted then and there by immediately turning away from coal and generating more electricity from nuclear plant. In short, some of the familiar features of nuclear power which make for its very great inflexibility, namely its capital intensity and lead time, make it a very clumsy way to protect electricity consumers against shocks from unexpected increases in fuel prices and shortages of fuel.

The story does not end here, however, because it must be remembered that even the CEGB is fallible. Suppose that the kind of shocks which it so confidently predicts in its case for the Sizewell PWR do not actually happen. The CEGB has then invested in a programme of nuclear stations, which can be assumed to have higher generating costs than coal plants, in order to acquire a fuel diversity which it does not need. The cost of such an error will be high for the reasons so often discussed before. There can be no amelioration by shifting to other fuels, the infrastructure of nuclear power must be maintained and the base load is crowded out making the next round of investment even less favourable to nuclear power, even though the performance of new designs may be a considerable improvement over the old. The costs of an erroneous investment in nuclear power will be very high, and this is as true if they are supposed to provide cheap electricity as if they are supposed to protect against shocks from high fuel prices and shortages.

Another kind of error which would prove similarly expensive arises if the price increases and shortages foreseen by the CEGB really do happen, if nuclear plant is built to counter this, and if it later turns out that other ways of achieving fuel diversity would have been cheaper than nuclear power. Suppose that much to everybody's surprise, oil prices fall to a level where, although dearer than coal, it is worth burning to achieve fuel diversity and that electricity from this source is cheaper than that from nuclear power. Or suppose that electricity from renewable energy sources turns out to be cheaper and more plentiful than expected, enabling renewables to be used to obtain diversity of fuel supplies. In either case, the CEGB would regret having built its nuclear plant because there were now cheaper ways of achieving fuel diversity, but there is nothing that can be done because of the capital intensity of the nuclear plant. As before, such erroneous investment would be very costly. The CEGB's error is the classical one of basing decisions on the best guesses which can be made today, in this case that oil continues to be expensive for fifty years and that renewable energy sources cannot meet more than 20 per cent of the CEGB's fuel needs by this date. If fuel diversity is required, then these beliefs entail that nuclear plant is the best way of achieving it. But this is altogether the wrong way to approach the uncertainties in a decision. Even the CEGB admits that the uncertainties here are very large, and so the problem is not to select the best way of achieving fuel diversity given today's best guesses about the future; instead the CEGB should seek to develop a generating system which can respond to whatever the future holds, to whatever happens to coal and oil prices and to whatever scale renewable energy turns out to be useable. Such a flexible generating system has no place for nuclear power.

A final problem in achieving fuel diversity from nuclear power is that many nuclear stations are needed. There is a sleight of hand here, for the CEGB argues for the PWR at Sizewell on the grounds of fuel diversity, and yet has to admit that its impact on the Board's reliance on coal will be minimal, a saving of only 2.5 million tonnes per year. Fuel diversity is a benefit from the Sizewell reactor only if it is part of a much larger nuclear programme (Electricity Consumers' Council (1983)). If this is the case, then the familiar problems about piecemeal and serial ordering of nuclear plant arise. If reactors are ordered serially, then the benefits of fuel diversity will be achieved quickly, if they are real. If, however, the kind of shocks nuclear power is supposed to protect against do not happen, or if they do happen but cheaper ways of achieving fuel diversity than nuclear power are possible, then serial ordering compounds the cost of error. The reverse is true of piecemeal ordering. Piecemeal ordering ensures that erroneous investment in nuclear plant is limited, but at the cost of greatly delaying the benefits from fuel diversity, should it be needed and should nuclear power be the best way of achieving it. Thus the familiar features of nuclear technology ensure that there is a problem over how units should be ordered, whether these units are needed to provide cheap electricity or diversity of fuel supplies.

What has been shown in this section is that the properties of nuclear technology which make it, when considered alone, highly inflexible also make it, when considered as part of a whole generating system, a very clumsy way of protecting electricity consumers from shocks from sudden and unexpected increases in fossil fuel prices and fossil fuel shortages. The original assessment of the technology's flexibility is not altered as the system under consideration expands from the nuts and bolts of the technology itself to the technology within a generating system. The CEGB is simply wrong in claiming that nuclear power would confer flexibility on their generating system by diversifying fuel supplies. This result confirms the hypothesis put forward earlier in this chapter that flexibility cannot increase as the boundary of a system is expanded. If this is the case, then adding considerations of flexibility to the usual prescriptions of incrementalism does not provide a back door route for the re-entry of the boundary problem which causes so many problems for synoptic rationality.

References

CEGB (1982a), *Central Electricity Generating Board, Proof of Evidence, Sizewell 'B' Power Station Public Inquiry, Proof 1, CEGB Policy*, CEGB, November.
CEGB (1982b), *Central Electricity Generating Board, Proof of Evidence, Sizewell 'B' Power Station Public Inquiry, Proof 4, The Need for Sizewell 'B'*, CEGB, November.
Electricity Consumers' Council (1983), *Statement of Case to the Sizewell 'B' Power Station Public Inquiry*, London.

PART IV

CONCLUSIONS

13   CONCLUSIONS

The conclusions drawn from this study may now be briefly presented. For the most part this chapter merely draws together conclusions which have been stated previously.

1. Theories of policy making

The study of nuclear power provides support for Lindblom's partisan mutual adjustment. Decisions about nuclear technology cannot be made in accordance with the prescriptions of partisan mutual adjustment, which would amount to a falsification of the theory if such decisions were like those about any other technology. But they are not. It has been shown, quite independently of partisan mutual adjustment, that nuclear power is a far from ordinary technology. Its very great inflexibility makes planning the technology an exercise which is very open to error, and also ensures that whatever mistakes are made are likely to be expensive. Nuclear power is therefore very different from other new technologies, and partisan mutual adjustment deserves credit for revealing this to us. My own account of social choice, critical decision theory, is similarly supported by the study of nuclear technology. This theory entails partisan mutual adjustment in an amended form where flexibility is an important requirement of the choices made by partisans. The key role of flexibility is well illustrated by the troubles and failures which form such a prominent part of the history of nuclear technology.

Similar tests of partisan mutual adjustment and critical decision theory may be imagined. Some item should be identified which cannot be accommodated within the rules laid down by these theories. If, on closer inspection, the item appears perfectly ordinary and trouble free, then this amounts to a falsification of the theories because they cannot allow for decisions about such an everyday item. If, on the other hand, it is found that the item involves peculiarly high risks, then this will corroborate the theories. Candidates here might include fusion reactors and large, integrated transport schemes.

A second area for research is the measurement of flexibility, as discussed in chapter 11. There is a very wide ranging, rich and varied literature which is ripe for exploitation. It is sure to yield valuable results about the making of tactical decisions on technology.

2. The control of technology

The chief lesson of this book is that technologies vary in the ease by which they may be controlled through the normal policy making machinery, and that technologies which are difficult to control in this way ought to be avoided. Technologies such as nuclear power, which cannot be controlled by the usual political adjustment of partisans, are likely to impose heavy economic and social costs because whatever ill effects they may prove to have, there is little that can be done to alleviate them. It is people who have to adjust to the technology, not the technology to people. In assessing a technology at a strategic level it is therefore vitally important to discover whether it is flexible enough to be brought under proper political control. Any technology assessment done in the future therefore ought to consider the flexibility of the technology being analysed as a central question in the way developed here, especially in chapter 9, which used the breeder reactor as a model example. In its early days, there were high hopes that technology assessment would be able to provide a synoptic overview of all aspects of a proposed technical change, social as well as economic, covering higher as well as first order effects. Not surprisingly, these hopes soon came to nothing, and there has since been a welcome move towards conducting assessments in ways which are more relevant to incremental policy making (Menkes (1981)). An important step in continuing this maturing process would be the consideration of flexibility in assessments of technology.

A second conclusion under this head is that there is a clear need for a much better understanding of the problem of scale in planning technological projects because the weakening of political control may often be attributed to the very large scale on which so many technologies are developed. It has been shown in chapter 11, for example, that LWRs in the United States were ordered in a unit size which ought to have been clearly recognised by all concerned as too large. This may be an extreme case, only further research will tell, but it illustrates a gross insensitivity to the problems of scale which can hardly be unique to that country's electricity supply industry. It seems clear that there is valuable research to be done on the general problem of scale. This was, in fact, started by the International Institute for Applied Systems Analysis, but fell victim to a tightening of research funds (Tomlinson (1977, 1982)). It is in urgent need of revival, there or elsewhere.

3. The role of government in promoting technology

Burn has argued that nuclear power has been developed more successfully in the United States than in Britain. This he attributes to the active encouragement in America of competition between many reactor types and between manufacturers, compared to the narrow front R & D and constant government involvement which marks the British story. This thesis has been rejected. The history of nuclear technology in the United States, of which Burn paints a very biased picture, tells very much the same story as its development in Britain. Reactors were ordered prematurely, in a serial fashion, in units which were too large, with the benefit of very little operating experience, and the consequent errors in investment have proved to be very costly. At the bottom of troubles in Britain and in America are the familiar features of nuclear power which make it so inflexible, its high capital cost and capital intensity, long lead time, large unit size and dependence on infrastructure. These have combined to cause many costly errors in both countries. This can explain the similar failure of nuclear power in the two countries, despite the very different political arrangements which were made for the new technology. No appeal needs to be made to the different political systems to explain the great difference in the technology's performance in the two countries, because this difference in performance is an illusion of Burn's. Burn's deepest error, though one shared with all other commentators to date, is to believe that nuclear power is an ordinary kind of technology. If this were so, then it might be expected that learning about the innovation would be prompted by the kind of competition encouraged in the United States. We now know, however, that nuclear power is a far from ordinary technology. It has properties which make learning slow and expensive, even in a system as large as the United States. Learning might have been slower in Britain because of the narrow front taken for reactor development and because of government involvement in the technology, but the difference cannot be more than marginal. At the root of slow learning is the technology itself, not the political philosophy under which it has been developed.

It might be thought that learning about different reactor designs in a system as large as that of the United States would be easy, but a superficial inspection reveals the problems of learning for smaller electricity systems, such as those possessed by Britain and France. Burn says nothing about these special problems, but they are very important in understanding the history of nuclear power. The history of British nuclear power clearly reveals the problems of learning about an inflexible technology in a relatively small economy. The steady, routine ordering of standard nuclear plant which has been the aim of so many governments, the utilities and reactor suppliers ever since the first orders for Magnox plants, has been vitiated by the large unit size and long lead times of reactors and by an electricity demand which refuses to increase at anything above a modest rate. The Magnox reactors were ordered as a result of the 1955 White Paper, although it was soon learned that these were not competitive with coal. There were eventually nine stations with a total capacity of 4.4 GW. The next step along the nuclear road was the AGR, of which five were ordered with a capacity of 6.5 GW. These were greatly delayed in construction, but electricity demand fell dramatically below forecasts so they were not needed to meet demand. After this, reactor choice was discussed at agonising lengths, but continuing low rates of increase in electricity demand meant that matters were not urgent, although two more AGRs were ordered, largely to assist the nuclear construction industry. Export prospects which had never been very rosy, declined to zero as American LWRs became the favoured technology.

This is the entire British experience of nuclear power over the past twenty-eight years. The CEGB now wishes to learn about PWRs, and to this end is seeking permission to build one at Sizewell in England. If their attempt is successful, the reactor would be operational around 1995, to mark the fortieth annniversary of the White Paper whose publication marked the dawn of the nuclear age. This single reactor will give little information about capital costs, because there is a considerable scatter of costs, even for very similar plants. Nor will it yield a great deal of knowledge about the PWR's ability to be built sufficiently cheaply in accordance with British standards of safety. Once it is working, all it signifies is that ten years or more ago the PWR design was acceptable to the British standards of that day. These may, of course, have changed in the meantime because of changes in the British perception of safety or from operating experience elsewhere in the world. Thus, after forty years the British experience of reactors will still be depressingly small, and the great programme of steady, routine orders of standard reactors will remain as much a dream as it ever was. At the bottom of this is the nature of the technology itself, with its large capital costs and capital intensity, long lead time and large unit size. Technology of this sort is bound to be developed in a very sporadic way in an economy the size of Britain's, especially one which shows only a modestly increasing demand for electricity and cannot capture overseas orders. To be more precise, its development is bound to be sporadic unless some very special arrangements are made for it. This takes us on to the history of nuclear power in France, where just such arrangements have been made.

In France, the normal political system has been radically adjusted to allow the rapid and smooth development of nuclear power. One reactor design was chosen, at a time when there was no commercial nuclear plant of any kind to provide experience on which such a decision might be based. A monopoly supplier of reactors was established, with a guarantee of steady orders to permit the production of many standard units per year. To prevent electricity demand from becoming an obstacle to the programme, it has been manipulated by favourable tariffs. This has made the French nuclear programme a great success, from a narrowly technological point of view. At a political and economic level, however, the costs are profound. Electricity is used where other fuels would have been cheaper and there are growing complaints that Framatome is exploiting its monopoly in reactor supply, which is said to be adding something like 10 per cent to capital costs (Parliamentary Committee on Production and Trade (1983)). The financial strain is beginning to tell on EDF which lost 8 fr. billion last year and expects similar losses this year despite two increases in tariffs and a cost cutting programme. Beyond these immediate problems is the risk of depending on one reactor design for such a large fraction of the country's primary energy supply. If design faults are revealed by future operating experience, this would be hugely expensive because of the capital intensity of the plant. This is a reflection of perhaps the deepest problem of them all, the complete lack of political control over the nuclear programme. Whatever problems the reactors produce, there is simply no question of reversing the programme; this is an option which was closed a long time ago, thanks to the technology and the degree of commitment necessary to exploit it.

A further point raised under this head is the role of government in the promotion of technology. A common argument for government support of new technology is that the risk is too great to attract private funds. From his study of the prototype breeder reactor in West Germany, Keck (1980, 1981a, 1981b) argues that government involvement is likely to lead to a poor assessment of the technology's economics. A better view is to be expected from private firms with a large financial commitment to the technology, though this may sometimes usefully be shared by government. Keck also attacks a second argument for government support, that the costs involved may be so large as to be beyond the pocket of any private firm. He shows that in many nuclear projects the sums spent on R & D, though large, are no greater than those spent on R & D by private firms standing alone. There is, however, a third argument which Keck does not discuss, that the technology may take such a long time to deliver whatever benefits it may have as to make it unattractive to private capital.

This is often said about the breeder. It is such a long-term project that it could never hope to attract private funds, but it is nevertheless a technology which might be very valuable in the future. But when the importance of flexibility is recognised, it will be seen that a technology with a development time as great as the breeder's is bad technology. It lays a cloud of uncertainty over the planning of related technologies for many decades, a problem which is worsened in the case of the breeder by its dependence on infrastructure. Should existing reactors be built if only to increase stocks of plutonium for the breeder? If thermal reactors are going to be ordered, should we ensure that the one producing the most plutonium is chosen? Should reprocessing plants for existing reactors be built, if only to give experience for the eventual reprocessing of breeder fuel? How much effort ought to be put into energy conservation, or the development of alternative energy, bearing in mind that the breeder may be available to meet supply problems for ever? The answers to these questions, and many more besides, are influenced by the breeder. If it were certain that the breeder was going to be providing cheap electricity safely any time it was chosen to commercialise it, then the answers to all of these questions would be influenced one way. If, on the contrary, it was known that breeder technology was a failure, then they would be answered in another way. The problem, of course, is that the status of the breeder is not going to be known for decades. So for all these years, uncertainty hangs over all these questions. This raises huge problems for energy planning over a very long time. For this reason alone, and forgetting for a moment all the other problems caused by the breeder's inflexibility which have been discussed earlier, the breeder reactor is bad technology, which should receive neither private nor government support. At a general level, it may be said that technology with a very long development time is bad technology. This means that the third argument for government support of new technology goes the same way as the other two. If the development time is so long as to make the technology unattractive to private investors, then the technology is one which does not deserve development anyway.

The story of nuclear power must give rise to great reservations about government support of technology. Wherever government steps in to support its favoured infant, disaster is soon to follow. In the United States, the USAEC legitimated the wild promises made by reactor suppliers in the early boom. In Britain, the government insisted that the electricty utilities purchase Magnox reactors and arranged the disastrous AGR programme. The French nuclear programme is completely dependent on government support, which has had such a heavy political cost and may yet have an equally large economic one. Only the support of the French government makes EDF bold enough to claim a great success in the form of the world's first large breeder reactor whose electricity will cost

2.2 times that from a PWR, and which is being built at a time
when uranium supplies seem to be adequate for decades to
come. The security and comfort afforded by the British tax-
payer allow UKAEA engineers to welcome yet another postponement of the British breeder to 2020 or beyond as giving
them extra time to get it right when it is needed. The United
States has perhaps the least need of any major country for
the breeder, and indeed the American breeder has attracted no
substantial funds from utilities or vendors. Nevertheless, scientists and engineers in government employment are pushing for
continued work on the Clinch River Project funded by their
employer, perhaps to the tune of $3,200 million.

4. The future of nuclear power

The chief conclusion concerning nuclear power is that it is
a poor technology; it has features which make mistakes in its
planning both likely and expensive. To put it another way,
it is a technology which is not open to the normal process of
political control through partisan mutual adjustment, but it
is also a technology which already exists. What conclusions
may be drawn about its future? In the United States, the
nuclear industry is at a very low ebb and has received only
cancellations since 1978. The accident at Three Mile Island
and other incidents have reduced public confidence in the
technology and any revival of its fortunes is likely to be slow.
The French programme, on the other hand, amounts to an
irreversible commitment to nuclear power, however views of
the technology may alter. Between these extremes are countries such as Austria, the Netherlands, Sweden and West Germany, whose nuclear programmes are in the process of
reappraisal, in which the view of nuclear power developed here
might have a role to play.

In Britain the CEGB is seeking permission to build a PWR
at Sizewell which is now the subject of public inquiry, and so
a review of the country's commitment to nuclear technology is
timely. The problems of developing a large nuclear programme
without benefit of exports in a country as small as the United
Kingdom have been discussed earlier. Ever since 1955, the
British nuclear industry has been trying to reach the mature
position of producing a regular series of standard reactors for

the home market and overseas wherever export orders could be captured. This happy state is, however, still a dream, and will remain so unless the industry and government face up to the problems of developing nuclear power in a country of this size. There is a clear choice for Britain between continuing with no nuclear industry at all, or one which functions in the desultory way of the past or else learning from the French how to adjust the political and economic machinery so that it can accommodate a nuclear programme on the same impressive scale as their own. It must be recognised that there is a price to pay for a country the size of Britain acquiring a mature nuclear industry, and that is the kind of centralisation which has occurred in France. The French have shown the way: one reactor type, one standard design, one manufacturer, a commitment to steady ordering for many years ahead and the manipulation of electricity demand so that it does not slow the programme. The British way of intermittent ordering in penny packets shared out between a handful of consortia, with constant agonising over reactor design, and with orders waiting on increases in electricity demand is a recipe for a nuclear industry of the sort it has had in the past.

Is the French road a real possibility for the British nuclear industry? It hardly seems so. First of all, the pattern of energy supplies is vastly different. Britain has its own substantial reserves of coal, gas and oil where France has to import practically all of its fossil fuels. Secondly, nuclear power has become a political issue in Britain in a way in which it has not done in France. Finally, it is simply very hard to imagine the British political system tolerating the strains of centralisation in the way the French system seems to have done quite happily. The real choice for Britain is therefore between having a nuclear industry tottering along as it has done up to now, or else having no nuclear industry at all. This is the real issue which ought to be on the current British political agenda.

If it is decided to build more nuclear plants in whatever country, then considerations of flexibility should enter into the tactical choice of reactor type, size and so on. This strongly favours existing reactor designs over new ones. There is so much to learn about a particular reactor, and learning is so slow that what has been learned about today's LWRs and gas-graphite reactors must be regarded as precious. To start a whole new reactor design would be to invite the revisitation of all the troubles which befell the early LWRs in the United States and gas-graphite reactors in Britain. Reactors, of whatever design finally chosen, should be built in much smaller units than presently regarded as economic, for reasons explained in chapter 11. Building in small units speeds learning about the reactors' safety and economics, not least about what their true economies of scale are. Similarly, reactors should be favoured which are less capital intensive, and less dependent on infrastructure, for these are ways of enhancing the technology's very low level of flexibility.

Finally, it is clear from this study that there is absolutely no case for continuing R & D on the breeder. This is likely to be even more troublesome than today's reactors have been because of the breeder's higher capital cost, higher capital intensity, longer lead time, larger unit size and greater dependence on infrastructure. As the R & D which has been done has proceeded, the flexibility of the breeder has been consistently diminished from its already low level. It is a technology which is certain to be plagued by costly errors, since mistakes in planning a technology of such miserably low flexibility are both likely and expensive. The breeder will be a technology even less suited to the normal processes of political control through partisan mutual adjustment than existing thermal reactors. If this were not enough, the long development time of the technology casts a cloud of uncertainty over a large part of energy planning, as indicated earlier. The breeder is bad technology which ought to be avoided. It is irrelevant to the real problems of energy policy, which are the design and provision of energy systems which are flexible so that they can easily adjust themselves to whatever changes and shocks the future might bring. Breeder reactors have no place in such a system; their existence immediately reduces flexibility and makes the energy system more difficult to control.

References

Keck, O. (1980), 'The West German Fast Breeder Programme', *Energy Policy*, **8**, 277–92.

Keck, O. (1981a), *Policy Making in a Nuclear Programme, The Case of the West German Fast Breeder Reactor*, Lexington Books, Lexington, Mass.

Keck, O. (1981b), 'Fast Breeder Reactors: Can We Learn from Experience', *Nature*, **294**, 205-8.

Menkes, J. (1981), 'The Contribution of Technology Assessment to the Decision Making Process', *Technological Forecasting and Social Change*, **19**, 45-53.

Parliamentary Committee on Production and Trade (France) (1983), *The Policy on Public Orders in the Energy Sector*, Paris.

Tomlinson, R. (1977), 'OR and Systems Analysis—Practice to Precept', *Philosophical Transactions of the Royal Society*, **A287**, 355-71.

Tomlinson, R. (1982), 'Developing a Systems Approach to Problems of Industry—Lessons from IIASA', *Journal of the Operations Research Society*, **33**, 1-19.

APPENDIX ALL YOU NEED TO KNOW ABOUT NUCLEAR POWER

The arguments presented in this work can be understood with only the most passing acquaintance with the details of nuclear technology, which this appendix aims to provide. For those who wish to delve more deeply, there are innumerable books and encyclopedia entries which are so readily available as to require no reference here.

The reactors discussed are of two kinds, *thermal* and *breeder* reactors. All commercial reactors so far have been thermal ones, but the French expect to complete their commercial breeder, the Superphenix, in 1984 and construction to date has proceeded remarkably smoothly. The smaller American demonstration breeder reactor at Clinch River is much less advanced and at the time of writing is under heavy attack in Congress. Other breeder reactors around the world are of smaller scale and are for R & D purposes rather than the commercial generation of electricity. It seems sensible, therefore, to begin by describing the principles of thermal reactors.

A thermal reactor produces energy from the breaking apart, or fission, of atoms of uranium. Not all uranium atoms undergo fission, however, but only those of the isotope uranium 235, which makes up about 0.7 per cent of naturally occurring uranium. When an atom of uranium 235 breaks up it releases energy and produces atoms of lighter elements, called fission products, and several neutrons. The neutrons are emitted at high speed, but if they are slowed sufficiently by being made to pass through water or graphite, called moderators, the neutrons can cause further atoms of uranium 235 to break up. Neutrons slowed in this way are called thermal neutrons, hence the name thermal reactors. Consider an atom of uranium 235 in a reactor which breaks up to emit three fast neutrons. Two of these neutrons may be lost, by escaping from the reactor or by being absorbed by various atoms, leaving one neutron which, when slowed by a moderator, can bring about the fission of another uranium 235 atom. If this atom breaks up to give just enough neutrons to cause one more uranium 235 atom to undergo fission, then the fission reaction can continue indefinitely. A *chain reaction* exists and the reactor is said to be *critical*. If not enough neutrons are produced in the reactor to sustain the reaction, then it slows to a halt.

The number of neutrons in the reactor is adjusted by the movement of control rods, which contain a powerful neutron absorber such as boron. If the rods are lowered into the core of the reactor, where the fission reaction occurs, they absorb neutrons and slow the reaction. If they are raised, the number of neutrons in the core, and hence the rate of the fission reaction, increases. Heat produced in the reactor core is generally removed by carbon dioxide or water, which is then passed through a heat exchanger to raise steam which is used to generate electricity in the conventional way. Problems of radioactivity leakage prevent the carbon dioxide or water heated in the core from being used directly for electricity generation.

A thermal reactor therefore has the following major components: uranium fuel, a moderator to slow neutrons, control rods which absorb excess neutrons and a coolant to take away the heat produced in the core on its way to make electricity. Radiation is, of course, hazardous to health and life, and so the reactor with all the equipment associated with it which will become radioactive in normal operation is enclosed in a thick biological shield. Figure A1 is a diagram of a Magnox reactor, the first kind built for commercial operation in the United Kingdom. The name comes from a special alloy used to encase the uranium fuel. The Magnox reactor's fuel is ordinary uranium metal arranged in rods and inserted into a large block of graphite which acts as the moderator in this design. Cooling is done by blowing carbon dioxide through the core under pressure. The hot gas is then used to raise steam which is taken off in the usual way to produce electricity. The second-generation British design, the advanced gas cooled reactors (AGRs), were a development from this and continued to use graphite and carbon dioxide. An important difference, however, was that the later reactors produced more heat from a given volume of graphite. Uranium metal would not stand up to the temperatures in the core of an AGR and so uranium oxide fuel was used. More important, the designers were not satisfied with the modest heat output from naturally occurring uranium, only 0.7 per cent of which can be used in

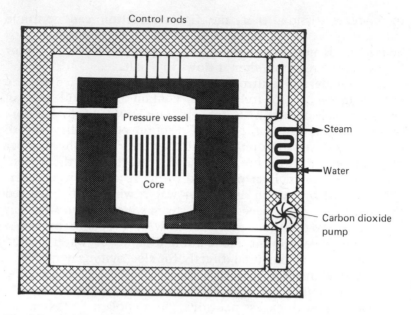

Fig. A1. Magnox reactor.

the fission process, and so they used uranium which had been *enriched* until it contained 2 to 3 per cent of uranium 235. In this way, these reactors rely upon the operation of very sophisticated uranium enrichment plants.

Most of the world's reactors are different, however, and only the British have persisted with this kind of technology. Two designs from the United States have come to dominate the world's reactors. Both use ordinary water, in the jargon of the industry this is *light water*, as a moderator and as a coolant combined, which makes for very compact reactors. They are therefore known as light water reactors (LWRs). In one design, the water is boiled to steam which is taken off, so that this is known as the boiling water reactor (BWR), while in reactors of the other type water is kept in a liquid state by maintaining a very high pressure. These reactors are pressurized water reactors (PWRs). Both types of reactor use enriched uranium fuel in the form of uranium oxide, so that their working is also dependent on enrichment plants.

Fuel in any kind of reactor cannot be left until all the uranium 235 in it has broken apart because the products of the chain reactor begin to interfere with the reaction itself. Fuel therefore has to be removed while there is still useable uranium 235 in it. If it is considered economic to do so, the fuel may be *reprocessed*, passed through a series of chemical reactions in a very carefully designed reprocessing plant to remove the remaining uranium. This leaves many radioactive products of the fission reaction to be disposed of in some way, and research is still underway on this problem. A third product of reprocessing is *plutonium*. Plutonium was once perhaps a quite common element in the earth, but it has long since decayed away until there is none occurring naturally. Thermal reactors therefore provide the only source of plutonium. Like uranium, plutonium has a number of isotopes which behave very differently, and one of these, plutonium 239, will undergo fission like uranium 235. This form of plutonium is made in all thermal reactors when an atom of uranium 238, which does not undergo fission, absorbs a neutron. Uranium 238 makes up nearly 99.3 per cent of natural uranium, the remainder being, of course, the isotope 235.

When fuel is reprocessed, useable uranium is returned to the reactor to produce more heat and so it is customary to talk of the *fuel cycle*. Uranium is mined, enriched, made into fuel and put into a reactor; after a time the fuel is removed and sent for reprocessing, which extracts unused uranium that is then returned to the reactor. There are two important by-products of the cycle. Enrichment to give a metal with 2 to 3 per cent uranium 235 obviously leaves a lot of uranium with virtually no uranium 235 in it. The process takes the uranium 235 from a large amount of natural uranium and squeezes it into a smaller amount, leaving a great deal of uranium which consists almost entirely of the isotope uranium 238. This so-called *depleted uranium* is quite useless for extracting energy because uranium 238 will not undergo fission. The other by-product is the plutonium from the reprocessing as shown in Figure A2.

Fuel from Magnox reactors has been routinely reprocessed for many years, but reprocessing fuels from other reactors has proved to be very much more difficult and expensive than once expected, so much so that thought is now being given to disposing of spent fuel without reprocessing. This would mean throwing away some useable uranium, but the cost of extracting it

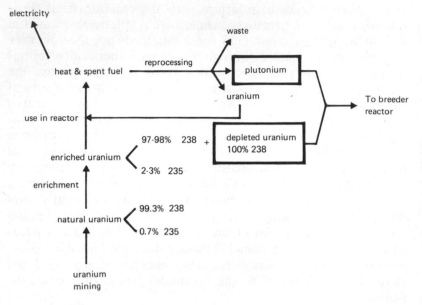

Fig. A2. Thermal reactor fuel cycle.

might be more than the benefits from putting it back into a reactor. The point is that reprocessing is not essential for the operation of thermal reactors; if it is worth doing then it may be done, but it is a question of economics and not of the functioning of the reactors.

The conception of the breeder reactor dates back to the early days of nuclear technology when the reprocessing of thermal fuel was seen as being routine. Thermal reactors produce two waste products, depleted uranium and plutonium which are quite useless, but the breeder reactor can turn these piles of useless metal into a virtually inexhaustible supply of energy. A breeder reactor has a very compact core and needs no moderator to slow down the neutrons it produces. It uses *fast* neutrons, hence the name fast breeder reactor. It can do this because its fuel is not uranium, but plutonium, or plutonium mixed with uranium. Fast neutrons cause the atoms of plutonium 239 to undergo fission, producing heat and more neutrons, which can go on to sustain a chain reaction. Up to this point the breeder resembles thermal reactors. The great difference is that the neutrons not only sustain a chain reaction, but they also combine with atoms of depleted uranium to produce plutonium 239. The depleted uranium may be mixed with the plutonium fuel or placed in a *blanket* around the core of the reactor where there will be a good supply of neutrons. The depleted uranium, which is useless as a fuel for thermal reactors, is therefore converted into plutonium, which can be extracted to provide further fuel for breeder reactors. Because of this conversion, the breeder can be designed so that it produces more plutonium than it consumes to make electricity, hence the name *breeder* reactor. This reactor therefore allows otherwise quite useless by-products of thermal reactors to be used to produce vast quantities of energy. In Britain alone, if the stockpile of depleted uranium could be used in this way, it would be equivalent to many billions of tonnes of coal.

The most studied design for commercial fast breeder reactors involves taking away the heat from the core by liquid sodium. These reactors are known as *liquid metal* fast breeders. The core of the reactor is very small, so much heat being developed in this small volume that a very efficient coolant, like liquid sodium, is needed. The sodium becomes radioactive and cannot be used directly to raise steam. Instead an intermediate heat exchanger passes the heat from the sodium to water, which then passes on to a further heat exchanger where steam is raised, as shown in Figure A3.

The breeder can only work, however, if the plutonium it makes can be separated so that it can be fabricated into new fuel. This requires reprocessing the spent fuel and the reactor's blanket; the reactor simply does not work unless reprocessing can be achieved, unlike thermal reactors where reprocessing is an option to be decided on economics. If reprocessing is technically possible, the rate at which a breeder produces plutonium is determined by the efficiency of the reprocessing plant, for even the best designed plants will lose some of the plutonium, and by the quantity of plutonium held up in store, transport, reprocessing and fuel fabrication—the so-called pipeline inventory. The time taken for one breeder to produce enough fresh plutonium to fuel a second reactor and to supply the pipeline inventory it will need is the (linear) *doubling time* of the reactor.

Breeder reactors of a commercial size will be considerably more expensive to build than existing thermal reactors, but they

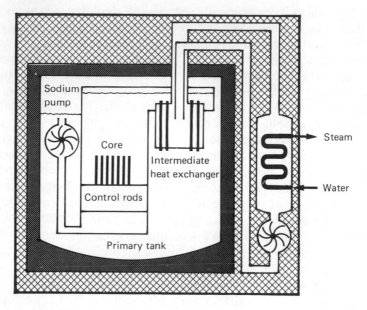

Fig. A3. Breeder reactor.

will be economic if uranium is very expensive. Thermal reactors can only use, at best, the 0.7 per cent of natural uranium which exists as uranium 235. A breeder, on the other hand, can turn the 99.3 per cent which is uranium 238 into a fuel, plutonium, so that it consumes uranium much more efficiently. At the moment uranium is cheap so there is no reason to build breeder reactors in order to use it efficiently, but if the price becomes sufficiently high in the future it will be economic to begin a programme of breeder reactors. The kind of world envisaged for the breeder is therefore one where coal, oil and natural gas prices are high and where uranium is also expensive, the more easily worked deposits having become exhausted as uranium is used to fuel thermal reactors. The rate at which breeder reactors can be built is, however, limited by the supply of plutonium. This can come from two sources, from thermal reactors or from breeders already in operation. In the early days of the breeder's development it was thought that they would have a doubling time of seven to twelve years, producing surplus plutonium at quite a rate. If this were to happen, then a breeder programme could soon stand on its own feet.

Early reactors would produce sufficient plutonium to fuel further reactors at such a rate as to enable the programme to grow, enabling breeder reactors to displace thermal ones quickly and be able to keep pace with modest increases in electricity demand. As more R & D has been done, however, the problems of achieving such a low doubling time have been revealed as much more serious than originally thought. Estimates for doubling time have lengthened to twenty, thirty, fifty and even sixty years, and there is even talk about the breeder which does not breed at all but just produces enough plutonium in its lifetime to fuel its replacement. This means that the number of breeder reactors will have to grow very slowly unless plutonium can be provided from thermal reactors. In this case, thermal reactors will have to be used alongside breeders and it will be many decades before they can be displaced by breeders.

INDEX

Abdulkarim, A. and N. Lucas 195, 209
Advanced gas reactor (AGR)
  description 241-2
  economies of scale 87
  errors in planning, cost of 85, 88-91
  errors in planning, likelihood 87-8
  forecasts of electricity demand 84-5
  forecasts of generating costs 82-5
  forecasts of load factor and capital
    costs 82-5
  infrastructure 84, 89-90
  lead time 85, 87-9
  serial and piecemeal ordering 89-91
  size 63, 88
  standardisation 86
Airport planning 131-3
Alexander, A. and D. Rice 15, 21
Armstrong, M. 131, 162
Ascher, W. 78, 91
Atomic Power Construction (APC) 83,
  198

Ball, D. 197, 209
Babcock and Wilcox 95
Bandwagon market for LWRs 95-7
Banfield, E. 3, 21
Bias in investment 66-8, 81-2, 89-90,
  99, 107
Boiling water reactor (BWR) 83, 88, 93,
  242 (see also Light water
  reactor)
Boundary problem
  and incrementalism 211-12
  and synoptic rationality 14, 212-8
  and uranium supply 13-14
Braun, E. and S. Macdonald 122, 124
Braybrooke, D. and C. Lindblom, 3, 21,
  40
Breeder reactor
  Clinch River 129, 236
  comparison with thermal reactors
    non-incremental 9-10, 136-7
  conditions favouring 129-30, 244-7
  cost
    direct and indirect 140, 144,
     155-6
    elements of 137-9
  cost benefit analysis 3-21

dependence on thermal reactors
  160-2, 244-7
description 244-7
development cost 4, 15
doubling time
  defined 138, 245
  factors determining 138-9
  forecasts of 143-4
  lengthening 156-62, 245-7
Dounreay Fast Breeder 128
economies of scale 11-12
effect on other energy decisions
  234-5
entrenchment 16-21
errors in planning
  cost 144-6
  likelihood 140-4
flexibility 140, 156-62
flexing in decision on 186-9
forecasts of electricity demand 10-11,
  141
forecasts of fuel costs 144
forecasts of generating costs 141
forecasts of load factor and capital
  costs 11-13, 141
forecasts of uranium price 13-15,
  141, 151-4, 165-9
French programme 59, 109-15,
  235-6
fuel cycle 244-7
future of 236-9
Government promotion of 234-6
hedging in decisions on 170, 186-9
infrastructure 140, 141, 145, 156-
  62
lead time 140-1
learning curve 142-3
out of reactor time
  defined 138
  slow learning about 143-4
Prototype Fast Reactor 128
reprocessing efficiency 138-9
reprocessing plant
  capital intensive 144-6
  economies of scale 146
safety and capital costs 143
serial and piecemeal ordering 146-50
size 140, 142-3

Breeder reactor (*cont.*)
　standardisation 11–12
　Superphenix 110, 129, 236–7
　timing of investment 164–9
　UK programme 128–30
　United States programme 1–21, 129, 235–6
　uranium supply 164–9
　W. German programme 234
Buckley, C., G. Mackerron and A. Surrey 152, 163, 168, 190
Bupp, I. and J.-C. Derian 58, 72, 93, 107
Burn, D. 58, 72–3, 85–6, 91, 93, 101, 104–7, 115, 231–2

Calder Hall 74–5
Camhis, M. 50–1
CANDU reactor 90, 194
Capenhurst enrichment plant 84
Capital cost forecasts—see load factor and capital cost forecasts
Carle, R. 111, 115
Carley, M. 3, 21
de Carmoy, G. 108, 115
Central Electricity Generating Board (CEGB)
　in AGR decisions 82–8
　case for Sizewell B 218–25, 236–7
　in Magnox decisions 73, 75, 77
　net effective cost calculations 62–5
　relationship to UKAEA 73, 75, 104
　substitution of coal by nuclear 66
　views on wave power 35
Chateau, B. and B. Lapillone 183, 189
Cheshire, J. and A. Surrey 183, 190
Chow, B. 138, 142, 163
Clinch River project 129, 236
Coal generating costs 62, 75–7, 79, 94
Collingridge, D. 3, 16, 19, 21, 38, 40, 47–8, 51, 55–6, 58, 92, 107, 109, 115, 134, 170, 182, 190, 195, 209
Collingridge, D. and J. Douglas 44, 50, 51
Collins, N. 208–9
Combustion Engineering 95
Commissariat a l'Énergie Atomique (CEA) 109–10, 112, 152
Commission Consultative pour la Production d'Électricité d'Origine Nucléaire 110
Committee for the Study of the Economics of Nuclear Electricity 65, 72, 77

Control cost 58
Corrective responsive time 57
Critical decision theory 47, 229–30

Dawson, F. 93, 107
de Carmoy, G. 107, 115
de Neufville, R. 132, 190
Department of Energy (UK)
　case for breeder 129–30
　energy conservation 139–40
Diablo Canyon 199
Dilemma of control 37–8
Disjointed incrementalism
　defined 26–30
　and nuclear power 31–40
Diseconomies of scale
　bias of future investment 197
　flow of earnings 197
　lead time 196
　learning about construction 196
　learning about performance 196
　in LWRs 199–207
　safety 196
　step effect 197–8
Dounreay Fast Reactor 128
Dungeness B Appraisal 83, 86, 88, 91

Economies of scale
　for AGR 87
　for breeder 11–12
　for LWR 101–2, 112, 119–207, 203–4
　for Magnox 75
Electricité de France (EDF)
　French nuclear programme 109–10, 235
　pricing policy 113–15
　reactor choice 112
Electricity Consumers' Council 224–5
Electricity demand forecasts
　in AGR decisions 84–5
　in breeder decisions 10–11, 141
　by EDF 112
　in hypothetical reactor decisions 61
　in LWR decisions 98
　in Magnox decisions 74–5, 78
Electricity supply, vulnerability 186–9
Energy conservation, hedging and flexing in 182–6
Energy forecasting, approaches to 182–6
Entrenchment 16–21, 37–40, 122
Errors, cost of 56, 57
Etzioni, A. 48–51
Eurochemic reprocessing plant 159

Evans, S. 208, 210

Fiering, M. 208–9
Finon, D. 107, 113, 115, 138, 142, 144, 159, 163
Fisher, J. 195, 209
Fission 204–1
Flexibility
  boundary problem 215–18
  concept of 212–14
  in critical decision theory 47, 229–30
  in decisions under ignorance 55–6
  enhanced by nuclear power 218–25
  Lindblom's views 47–8
  measurement 128, 133–4, 207–8, 217–18
  of nuclear power
    breeder 140, 156–62
    hypothetical reactor 55, 70–2
    LWR 98, 199–207, 218–25
Flexing (see also Hedging)
  benefits not ignored 185–6
  for breeder 170, 186–9
  definition of 170–3
  for energy policy 182–6
  for H-bomb 178–81
  for Magnox 174–8
  for MIRV 181–2
  revision of 174–8
  vicious circle avoided 182–5
  worst outcome avoided 178–81
Framatome 110, 112, 115
Franks, C. 58, 72
French nuclear programme 59, 109–15
Friedman, Y. and G. Reklaitis 208–9
Friedmann, J. and B. Hudson 3, 21
Fuel Cycle 244
Fuel diversity 218–25
Fumas, V. and A. Whinston 208–9
Fundamental and incremental decisions 48–50

Gandara, A. 93, 107
General Electric 93–4, 96
Generating costs—see Nuclear generating costs
Gershuny, J. 50, 51
Greenwood, T. 181, 190
Grinyer, P. and G. Wooler 208–9
Gross response time 57

Häfele, W. 148, 163
Hall, P. 121, 124
HAO and UP2 reprocessing plant 159

Hashimoto, T. *et al.* 208–9
Hedge (see also Flexing)
  benefits ignored 185–6
  for breeder 170, 186–9
  definition 170–3
  for energy policy 182–6
  for H-bomb 178–81
  for Magnox 174–8
  for MIRV 181–2
  revision of 174–8
  vicious circle encouraged 182–5
  worst outcome ambiguous 181–2
  worst outcome effected 178–81
High temperature gas reactor
  capital cost forecasts 10–13
  non-incremental comparison with other options 9, 139
  rival to breeder 5
  successor to LWRs 95
Hinton, C. (later Lord) 58, 72
House of Commons Select Committee on Energy 65, 72, 78, 91
Hydrogen bomb, hedging and flexing 178–81
Hypothetical reactor programme
  errors in planning, cost of 59, 66–72
  errors in planning, likelihood of 60–6
  flexibility 55, 70–2
  forecast of electricity demand 61
  forecast of generating costs 60–2
  forecasts of load factors and capital costs 61–3
  forecasts of uranium prices 61–2
  infrastructure 59, 66–7
  lead time 60–5, 68–70
  learning curve 61–2
  serial and piecemeal ordering 68–72

Ignorance, decisions under 55
Ince, M. 114–15
Incremental change (see Non-incremental change)
Incrementalism (see also Disjointed incrementalism and Partisan mutual adjustment)
  boundary problem 212–18
  decisions on nuclear power 23–40
  and synoptic rationality 118–19
Inflexible strategy
  breeder as example 136–62
  definition 134
Infrastructure
  AGR 89–90

Infrastructure (*cont.*)
  breeder 140-1, 145, 156-62
  hypothetical reactor 59, 66-7
  LWR 103-4
  Magnox 81-2
International Nuclear Fuel Cycle Evaluation 166

Jeffrey, J. 65, 72, 85, 91
Joint Committee on Atomic Energy (of US Congress) (JCAE) 58, 93

Keck, O. 142, 144, 163, 234, 239
Kerchner, O. 208-9
Komanoff, C. 93, 102-3, 107, 143, 163, 196, 202-4, 209

Lamont, N. 183
Leach, G. 183-5, 190
Lead pollution 8
Lead time
  AGR 85, 87-9
  breeder 140-1
  hypothetical reactor 60-5, 68-70
  LWR 100-4, 111, 199-200, 204-5
  Magnox 78-9, 80-2
Learning curve
  breeder 142-3
  hypothetical reactor 61-2
Lee, T. 195, 209
Light Water reactor (see also Boiling and Pressurised—water reactors)
  description 242-3
  economies of scale 101-2, 112, 203-4, 199-207
  errors in planning, costs of 99-100
  errors in planning, likelihood 98-9
  flexibility 98, 199-207, 218-25
  forecasts of electricity demand 98
  forecasts of generating costs 94, 111
  forecasts of load factors and capital costs 11-13, 94-7, 100-4
  forecasts of uranium prices 96
  infrastructure 103-4
  lead time 100-4, 111, 199-200, 204-5
  safety 106, 114
  safety and capital costs 103-4, 201-2
  serial and piecemeal ordering 99-100, 114-15
  size 95, 97, 112, 199-207
  standardisation 95-6, 111, 113-15, 204-5

Lindblom, C. (see also Disjointed incrementalism and Partisan mutual adjustment) 23, 25-6, 28-30, 40, 44, 48-9, 118, 123
Liquid metal fast breeder reactor (LMFBR)—see Breeder reactor
Load factor and capital cost forecasts
  AGR 82-5
  breeder 11-13, 141
  hypothetical reactor 61-3
  LWR 11-13, 94-7, 100-4
  Magnox 74-6
London motorway scheme 121
Lönroth, M. and W. Walker 114-15
Lovins, A. 183, 190
Lucas, N. 109, 113-14, 116

Magnox reactor
  description 241-2
  economies of scale 75
  errors in planning, cost of 80-2
  errors in planning, likelihood 75, 79
  forecasts of electricity demand 74-5, 78
  forecasts of generating costs 74-7
  forecasts of load factors and capital costs 74-6
  hedging and flexing in decisions on 174-8
  infrastructure 81-2
  lead time 78-9, 80-2
  serial and piecemeal ordering 80-1
  size 75, 79
Marschack, T. and R. Nelson 40, 208
Marshall, W. 147, 163
Marsham, T. 147, 163
Menkes, J. 230, 239
Merkhofer, M. 208, 210
Meyerson, M. 3, 21
Meyerson, M. and E. Banfield, 3, 21
MIRV, hedging and flexing in decisions on 181-2
Mitchell, J. 208, 210
Moderators 241
Monitoring 57, 171
Monitor's response time 57
Monopolies and Mergers Commission (UK) 62, 66, 72, 144, 163
Montenegro, J. 208, 210
Mooz, W. 93, 101-2, 108, 196, 199, 202, 204, 209

National Coal Board (NCB) 222

Net effective cost calculations 64–5
Nicholson, R. and A. Farmer 147, 151, 163, 170, 190
Non-incremental change
  breeder vs. thermal plant 9–10, 136–7
  coal vs. nuclear plant 9, 31–7
  definition 8
  problems caused by 122
  in strategic decisions 132–3
Nuclear generating cost forecasts
  AGR 82–5
  breeder 141
  hypothetical reactor 60–2
  LWR 94, 111
  Magnox 74–7
Nuclear Power Group, The (TNPG) 84

Organisation for Economic Co-operation and Development (OECD) 166, 168, 190
Out of reactor time
  definition 138
  slow learning about 143–4

Parker, Lord Justice 90, 91
Parliamentary Committee on Production and Trade (France) 233, 239
Partisan mutual adjustment
  airport planning 127–8
  and critical decision theory 47–8, 229–30
  is normative and descriptive 41–5
  and nuclear power 127–8, 229–30
  and nuclear power debate 31–7
  test for 41–50, 117–24
Patterson, W. 58, 72, 183–4, 187, 190
Perry, R. 93, 108
Piecemeal ordering—see Serial and piecemeal ordering
Planning errors in nuclear power, cost of
  AGR, 85, 88–91
  breeder 144–6
  hypothetical reactor 59, 66–72
  LWR 99–100
  Magnox 80–2
Planning errors in nuclear power, likelihood
  AGR 87–8
  breeder 140–4
  hypothetical reactor 60–6
  LWR 98–9
  Magnox 77–9

Plutonium
  British bomb programme 74
  credit 75
  production 243–4
  toxicity 16–21
Popper, Sir Karl 41–2
Pressurised water reactor (PWR) (see also Light water reactor) 93, 110, 112–13, 242–3
Proliferation of nuclear weapons 155–6
Prototype Fast Reactor 128
Pye, R. 208, 210

Quiles report 111

Radetzki, M. 152, 163
Reprocessing
  for breeder reactors 137–9, 143–4, 244–7
  for thermal reactors 90, 159–60, 242–4
Reversibility of nuclear investment
  AGR 89–90
  breeder 145–6
  low caused by capital intensity 39, 123
  and entrenchment 119
  hypothetical reactor 68–9
  LWR 103
  Magnox 80–2
Richels, R. and J. Plummer 142, 163
Rickover, Admiral 93
Rocky Flats fire 18, 20
Rolph, E. 93, 108
Rosenhead, J., M. Elton and S. Gupta 208, 210
Rosenhead, J. and S. Gupta 208, 210
Roskill, Lord Justice 131, 132, 190
Royal Commission on Environmental Pollution (UK) 148–9, 163, 187, 190
Rush, H. et al. 58, 72

Safety and nuclear capital costs
  breeder 143
  LWR 103–4, 201–2
Schapira, J. 109, 116, 159, 163
Scientific method 41–2
Serial and piecemeal ordering
  AGR 89–91
  breeder 146–50
  hypothetical reactor 68–72
  LWR 99–100, 114–15
  Magnox 80–1

Size of reactors (see also Diseconomies of—and Economies of—scale)
 AGR 63, 88
 breeder 140, 142–3
 LWR 95, 97, 112, 199–207
 Magnox 75, 79
Sizewell B 218–25, 232, 236
Smart, I. 198, 210
Smith, G. and D. May 50–1
South of Scotland Electricity Board (SSEB) 73, 84–5
Standardisation of reactors
 AGR 86
 breeder 11–12
 British hopes 232–4, 236–7
 LWR 95–6, 111, 113–15, 204–5
Stigler, G. 208, 210
Strategic choice
 and breeder 136–62
 defined 130–2
 role of flexibility 133–6
 statement of aims 130–2
Suez crisis 75
Superphenix 110, 129, 236–7
Surrey, J. and S. Thomas 62, 72, 197, 203, 205, 209
Sweet, C. 58, 72, 109, 111, 113–14, 116, 138, 159
Synoptic rationality
 and boundary problem 211–12
 consistency of decisions 23–4, 26
 criticism of 23–4, 28–9, 44, 117
 and nuclear power 3–21, 117–18

Tammen, R. 181, 190
Tactical choice
 between LWRs of different sizes 199–207
 definition 193
 role of flexibility 193–5
 trade off, flexibility and cost 194–5
 and scale diseconomies 195–9
 and scale economies 195
Thermal Oxide Reprocessing Plant (THORP) 90, 144
Three Mile Island, accident at 106
Tomlinson, R. 231, 239
Transistor 122
Truman, President 178

United Kingdom Atomic Energy Authority (UKAEA)
 AGR decisions 82
 breeder scenario 148–50
 functions 73
 narrow front R & D 59
 role in decision making 36–7, 104–7, 236
United States Arms Control and Disarmament Agency 181
United States Atomic Energy Commission (USAEC)
 cost benefit analysis of breeder 4, 7, 13, 17, 21–2, 137–8, 142, 147, 194
 functions 93, 95, 104–5
 H-bomb decision 178–9
 optimism about LWRs 58, 95–6, 98, 101, 235–6
United States Energy Research and Development Administration (ERDA) 138, 142, 164
United States National Academy of Sciences (NAS) 142, 147, 151, 163, 168, 190
Uranium
 concentration of supply 152
 enrichment 74, 84, 89, 153, 241–3
 exhaustion 165–9
 market complexity 150, 151–4
 price forecasts
  breeder reactor 13–15, 141, 151–4, 165–9
  hypothetical reactor 61–2
  LWR 96
 price history 152–3
 spot market 152

Van der Vet, R. 208, 210
Vaughan, R. and A. Farmer 147, 164

West Valley reprocessing plant 159
Westinghouse 93, 96, 110, 154
Williams, R. 58, 72, 73, 86, 91
Wonder, E. 58, 72
World Energy Conference 14, 22, 148, 164, 168, 190

York, H. 178, 181, 190